T0389101

Biorefineries

Biorefineries
Design and Analysis

Carlos Ariel Cardona Alzate
Department of Chemical Engineering Universidad Nacional de Colombia sede Manizales

Jonathan Moncada Botero
Department of Chemical Engineering Universidad Nacional de Colombia sede Manizales

Copernicus Institute of Sustainable Development Utrecht University

Valentina Aristizábal Marulanda
Department of Chemical Engineering Universidad Nacional de Colombia sede Manizales

CRC Press
Taylor & Francis Group
Boca Raton London New York

CRC Press is an imprint of the
Taylor & Francis Group, an **informa** business

CRC Press
Taylor & Francis Group
6000 Broken Sound Parkway NW, Suite 300
Boca Raton, FL 33487-2742

© 2019 by Taylor & Francis Group, LLC
CRC Press is an imprint of Taylor & Francis Group, an Informa business

No claim to original U.S. Government works

Printed on acid-free paper

International Standard Book Number-13: 978-1-138-08002-7 (Hardback)

Library of Congress Cataloging-in-Publication Data

Names: Cardona Alzate, Carlos Ariel, author. | Moncada Botero, Jonathan, author. | Aristizábal Marulanda, Valentina, author.
Title: Biorefineries : design and analysis / Carlos Ariel Cardona Alzate, Jonathan Moncada Botero, Valentina Aristizábal Marulanda.
Description: Boca Raton : CRC Press, 2019. | Includes bibliographical references and index.
Identifiers: LCCN 2018018742 | ISBN 9781138080027 (hardback : alk. paper) | ISBN 9781315114088 (ebook)
Subjects: LCSH: Biomass conversion—Environmental aspects. | Biomass conversion—Economic aspects. | Biomass conversion—Social aspects. | Biomass—Refining.
Classification: LCC TP339 .C369 2019 | DDC 662/.88--dc23
LC record available at https://lccn.loc.gov/2018018742

Visit the Taylor & Francis Web site at
http://www.taylorandfrancis.com

and the CRC Press Web site at
http://www.crcpress.com

To my mother… forever an angel. To Alexei, Jean Paul, and Libia…

deserving all my love. To those students who are always my friends.

Carlos Ariel Cardona Alzate

To my family and friends.

Jonathan Moncada Botero

To Amparo, Isabela, Juancho, and Juan Camilo for their love,

encouragement, and support in all aspects of my life.

Valentina Aristizábal Marulanda

Contents

Preface

The biological resources, and in the past few decades the different types of biomass, are being considered as an attractive raw material for obtaining a range of value-added products that could be the same or similar to those that nonrenewable oil is providing today. However, some strategies are needed to demonstrate that biomass is really a sustainable option to replace oil and other non-biodegradable feedstocks. The first need is to understand the biomass as any multicomponent organic matter that can be used as energy or any value-added product source. Biomass is not only a very old energy carrier based on the sun but also an important natural raw material that in the case of wood has being historically burned to heat homes or to cook food. So, biomass itself can be considered finally a feedstock if some transformation is applied in a very well-designed way. Based on this, it is imperative before any product from biomass is obtained to consider integrally all the remaining biomass for other products. This task is the purpose of the best strategy we have today: the biorefinery concept. This concept is based on the fractionation of biomass into different products, where, according to the physical and chemical nature of raw materials (organic waste, energy crops, and aquatic biomass), the performance and costs can vary considerably.

The authors of this book introduce readers to all the concepts needed to design and analyze a biorefinery as a real sustainable strategy to use integrally the biomass. There are so many biomass raw materials, conversion routes, and technologies that the not totally well-known biomass can generate too many alternatives making a good design impossible without considering as maximum all these possibilities. This book tries to show one of the ways to achieve feasible configurations in the design of biorefineries.

Emphasis is placed on the heuristic approach and the simulation of the total biorefinery using mainly Aspen Plus. The entire book includes a discussion about technical, energy, economic, environmental, and social assessments of the biorefineries. The last element is possible only after the simulations, which is important to confirm the final expected sustainability in the designed biorefineries.

Chapter 1 describes an overview of the most relevant definitions of biorefinery and its conceptual evolution from stand-alone bioprocesses to the currently known robust integrated systems. Chapter 2 includes an overview of all bio-based feedstocks and their respective classification. Then, platforms such as articulating shaft in a biorefinery are indicated. Finally, the main products of the biorefineries are discussed and some examples are presented. Chapter 3 involves a heuristic approach related with the design of a biorefinery system using the hierarchy, sequencing, and integration concepts. Moreover, the hierarchy, sequencing, and integration concepts are explained and discussed using examples of biorefineries reported in the literature aiming to demonstrate the applicability of the design approach. Chapter 4 includes a detailed explanation of the process simulation, mass, and energy integration along with the techno-economic assessment of biorefineries. Chapter 5 describes a wide overview of methodological approaches to do an environmental assessment applied to biorefineries. Additionally, the more commonly used methodologies and some examples of applicability are described. Chapter 6 involves the first approaches to social assessment applied to biorefineries using mainly as example the Colombian case given the socioeconomic characteristics of this country. Throughout this chapter, the most relevant concepts are indicated to determine the social impact of biorefineries: job generation at land and process level, food supply, and health. Chapter 7 indicates some specific examples of biorefinery systems considered a logic result of their evolution from main biofuel factories to multiproduct complex biorefineries. Additionally, this chapter presents a general perspective of the logistic concept applied to biomass supply systems. Chapter 8 describes the indices to assess a biorefinery considering its complexity, conversion capacity, water and reagent consumption, and economic results. Chapter 9 finally introduces readers to key challenges for future development of biorefineries.

It is important to note that the authors were involved in all the chapters. However, even if this book shows more than 23 years of research in this type of topics under the leadership of Professor Cardona, the heart in writing and compilation of all theoretical developments of this book was Valentina Aristizábal Marulanda, who is ending her PhD studies in chemical engineering with an emphasis on biorefineries.

Acknowledgements

The authors believe that any book is just an open invitation to readers to discuss and contribute in any specific field. Biorefineries are a topic needing more and more perspectives of design and analysis to really be introduced to an industry as a sustainable way for biomass processing. The authors expect many contributions and new approaches from readers.

Finally, but being really the most important, the authors acknowledge many organizations and people who are the sponsors of the research and are presented below: Dirección de Investigación (DIMA) and the Dean of the Facultad de Ingenieria y Arquitectura (FIA) at Universidad Nacional de Colombia sede Manizales, Departamento Administrativo de Ciencia, Tecnología e Innovación (Colciencias) through call 727 of 2015 and the Government of the Caldas region in Colombia for their valuable financing support.

All the former and actual students of our research group in Procesos Químicos, Catalíticos y Biotecnológicos (PQCB) for being the heart of all we do and publish (not mentioning all): Luis Fernando Gutiérrez, Franz López, Luis Javier, John Alexander Posada, Julian Quintero, Kelly Dussán, Luis Eduardo Rincón, Laura Daza, Juan Camilo Carvajal, Ivonne Cerón, Alexandra Duarte, Mónica Valencia, Angela Gonzalez, Javier Dávila, Valentina Hernández, Yessica Chacón, Daniela Parra, Christian Botero, Mariana Ortiz, Estefanny Carmona, Paula Marín, Jhonny Poveda, Jimmy Martínez, and Jose Gonzalez, among others.

A special acknowledgment is conveyed to chemical engineers, Carlos Andrés García Velasquez, Juan Camilo Solarte Toro, and Sebastián Serna Loaiza for developing the biorefinery concepts by including thermochemical processes, catalytic routes for biomass transformation, and small-scale biorefineries respectively. In addition, they contributed hardly to Chapters 3 and 4 as well as to the overall revision of this book.

Other collaborators and friends in different projects and students exchanges are also recognized by the authors: Eulogio Castro Galiano from Universidad de Jaen, Moshe Rosenberg from the University of California, Davis, Ashwani Gupta from the University of Maryland, Mahmoud El-Halwalgi from Texas A&M, Evan Beach from the University of Yale, Agustin J. Castro Montoya from Universidad Michoacana de San Nicolás de Hidalgo, Alfredo Martínez Jimenez from Instituto de Biotecnologia UNAM, Carlos Eduardo Orrego Alzate from Instituto de Biotecnología y Agroindustria UNAL Manizales, Germán Aroca Arcaya from Pontificia Universidad Católica Valparaiso, Victor Haber Pérez from UENF-Brazil, Florbela Carvalheiro–Francisco Girio–Luis Duarte from Laboratorio Nacional de Energia y Geologia-Lisbon, Jose A. Teixeira from Universidade do Braga, Mercedes Ballesteros from

CIEMAT-Madrid, Eric Fraga from UCL-London, Annette Weidtmann from the University of Hohenheim, Konstantinos Moustakas from the National Technical University of Athens, Mariano Martín Martín from the University of Salamanca, and Jin suk Lee from the Korean Institute of Energy Research, and finally with very special mention professors Yuri Pisarenko and Leonid Serafimov from the MITXT Moscow.

Carlos Ariel Cardona Alzate
Jonathan Moncada Botero
Valentina Aristizábal Marulanda

Authors

Carlos Ariel Cardona Alzate is full professor of the department of chemical engineering in Universidad Nacional de Colombia sede Manizales since 1995. He received MSc and PhD degrees in chemical engineering from the Lomonosov Moscow State Academy of Fine Chemical Technology in 1995. From 1996 to 1997, he worked at Universidad de Caldas supporting a new program in food engineering. Dr. Cardona's research focuses on developing nonconventional separation processes, thermodynamics, integration processes, process engineering, biorefineries, climate change, and agroindustry. In particular, he has worked on different research projects concerning the chemical and biochemical process designs, biofuels research and development, and economic and sustainable utilization of Colombia agroindustrial wastes, among others. He has authored and coauthored more than 150 research papers as well as 12 research books and 44 book chapters. Additionally, he has presented more than 250 works at scientific events. Currently, he leads the research group in Procesos Químicos, Catalíticos y Biotecnológicos (PQCB) at Universidad Nacional de Colombia sede Manizales.

Jonathan Moncada Botero received BE in chemical engineering and MSc (*Laureate distinction*) degrees from Universidad Nacional de Colombia sede Manizales in 2010 and 2012, respectively. Since 2014, he has worked as a junior researcher, and simultaneously, he completed his PhD in sustainability assessment of biorefinery systems at Copernicus Institute for Sustainable Development, Utrecht University, the Netherlands. Moncada has scientific expertise in topics such as techno-economic analysis, sustainability assessments, biorefineries, process modeling and simulation, Aspen Plus, and life cycle assessment. In particular, he has worked on different research projects concerning the modeling and techno-economic and environmental assessment of processes and biorefineries. He has authored and coauthored more than 25 research papers as well as four book chapters. Additionally, he has presented more than 30 works at scientific events.

Valentina Aristizábal Marulanda received BE in chemical engineering and MSc degrees from Universidad Nacional de Colombia sede Manizales in 2013 and 2015, respectively. Aristizábal Marulanda's research focuses on processes engineering, process modeling and simulation, and design and analysis of biorefinery configurations considering technical, economic, environmental, and social aspects. She has worked in the production of jet fuel through a furfural platform and in the development of projects related to the valorization of residues. She has authored and coauthored more than six research papers as well as five book chapters. She has presented more than 33 works at scientific events. Aristizábal Marulanda has done internships at the University of Salamanca, Spain, and the University of Minho, Portugal. Currently, she is doctoral candidate in chemical engineering at Universidad Nacional de Colombia sede Manizales, and she was qualified as junior researcher in 2017 by the Departamento Administrativo de Ciencia, Tecnología e Innovación (Colciencias) in Colombia.

1

The Biorefinery Concept

This chapter aims to provide an overview of the most relevant definitions of biorefinery and its conceptual evolution from stand-alone bioprocesses to the currently known robust integrated systems. It also focuses on the examples of biorefineries at different development stages (e.g., commercial, demonstration plants, pilot plants, and R&D) to introduce the readers to think about a global perspective of their applications in the real world. A biorefinery processes the biomass integrally; thus, it takes into account upstream, midstream, and downstream operations. Following this reasoning, the main methods and techniques for pretreatment, conversion, and separation used in biorefinery systems are also discussed here. Additionally, as a biorefinery presents infinite possibilities in terms of routes and processing sequences, several strategies to design biorefineries, highlighting their pros and cons, are briefly described.

1.1 Biorefinery Concept

There are several points of view for biorefining leading to different definitions and positions. This section aims to describe, in a brief but concise way, the different perceptions about the biorefinery concept presented in the literature. The most important definitions for biorefineries and their evolution from stand-alone processes to integrated portfolios are described in this section.

According to the International Energy Agency (IEA) Bioenergy Task 42, a biorefinery is defined as "...the sustainable processing of biomass into a spectrum of marketable products (chemicals and materials) and energy (fuels, power and heat)" [1]. The systems that include a biorefinery can be described as a facility, a cluster, a spectrum, and a process. A biorefinery is the integrated processing of biomass to get a range of products based on recent advances in upstream, midstream, and downstream operations. For IEA's Task 42, a real biorefinery produces both energy and non-energy products.

According to the National Renewable Energy Laboratory and U.S. Department of Energy, a biorefinery is "...a facility that integrates biomass conversion processes and equipment to produce fuels, power, and value-added chemicals from biomass" and "...an overall concept of a promising plant where biomass feedstocks are converted and extracted into a spectrum of valuable products" [2,3].

Euroview defines biorefineries as "…integrated bio-based industries, using a variety of technologies to produce chemicals, biofuels, food and feed ingredients, biomaterials (including fibers) and power from biomass raw materials" [4].

Huang et al. (2008) defined biorefinery as processes that use bio-based resources such as agriculture or forest biomass to produce energy and a wide variety of chemicals and bio-based materials, similar to the modern petroleum refineries [6].

Bruins et al. (2012) defined the biorefinery concept as "Biorefineries use biomass to produce energy and chemicals, similar to petroleum refineries. They are being developed as a more sustainable alternative. However, although many of the end-products are similar, processing is very different" [7].

As a generalized idea, a biorefinery involves assessing and using a wide range of technologies to separate biomass into its principal constituents (carbohydrates, proteins, triglycerides, etc.), which can subsequently be transformed into value-added products in an economic, social, and environmental manner or the so-called sustainable way [7,8]. The main goals of the biorefineries are to maximize the value of the products obtained from the biomass, to increase competitiveness and prosperity of industry, to reduce the dependency on fossil fuels by producing the same or analogous substances demanded in the current market, to reduce the emission of greenhouse gases, and to stimulate regional and rural development [8].

A biorefinery is not totally a new concept because industries such as that of sugar, starch, and pulp and paper have been using technologies which are covered under a biorefinery concept [1,9,10]. Economic and environmental factors have been the drivers to guide these industries to its total development under the biorefinery concept. Specifically, factors such as global warming, energy supply, energy efficiency and costs, and agricultural policies, among others, have influenced these changes over the past few decades [11].

A clear example of the evolution of a bio-based industry is that of corn milling. This process began as a stand-alone processing technology to mill corn and isolate the starch, and later has been converted into a multiproduct process to also obtain streams such as gluten feed, gluten meal, kernel oil, and fiber [12,13]. This evolution clearly shows a multiproduct biorefinery path. This transformation was given in the 1970s, where initially the corn milling industry produced starch as the main product. The need to obtain value-added products based on the market demand, the oil crisis as well as the development of new technologies were the factors that motivated the production of starch derivatives such as glucose, maltose syrups, and high-fructose corn syrup (HFCS). Nowadays, these products represent the 37% of the production in an U.S. corn wet milling industry and are used as platforms to produce fuels and chemicals such as ethanol, citric acid, lactic acid, and gluconic acid via fermentation.

Many authors define the biorefinery concept as the analogy to that of oil refineries, where instead of using oil, biomass is fractionated into a suite of products [6,13,14]. When biomass and crude oil are compared, characteristics such as renewability, storage, substitutability, abundance, and carbon neutral (zero emissions) are highlighted, generating remarkable differences between these two raw materials [8].

Another main difference appears in the feedstock distributions and opportunities when biomass is fractionated into a family of products with different added values [6].

Two major elements make oil refineries and biorefineries different. The first are the raw materials themselves as those used in biorefineries have not experienced the biodegradation over millions of years to become crude oil. The second are the technologies themselves as those used for biorefineries are still under the development stage and due to complexity of the biomass itself as those become more complex to integrally and simultaneously produce bioproducts. The portfolio of products from a biorefinery not only includes those that are currently obtained in oil refineries but also many that cannot be obtained from crude oil such as food derivatives [6,13].

In a broad definition, biorefineries process a wide range of biomass types (e.g., organic residues, energy crops, lignocellulosics and algae biomass) to obtain several product streams (e.g., fuels, chemicals, power and heat, materials, and food and feed). It is important to note that the maximum use of the raw materials and minimum production of residues (increasing the number of products, if possible) is a clear sustainability goal of any biorefinery.

Generally, a biorefinery approach involves multiprocessing, in which the first step, following feedstock selection, typically involves the treatment of biomass. This step is conventionally referred to as pretreatment. After this stage, the biomass is pre-fractionated to access the main components for a further transformation through a combination of biological and/or chemical processes. The outputs from this step are generally named as platform chemicals (e.g., specialty chemicals or reducing sugars) that can be converted into chemical building blocks for further processing uses [15].

The products based on bioresources can be obtained in a single productive process (so-called stand-alone). However, the integrated production of chemicals, materials, energy, and food is probably a more efficient approach for the sustainable valorization of biomass toward a bio-based economy [16–18]. Future biorefineries will be able to mimic the energy efficiency of modern oil refining through heat integration and market development of co-products. Heat that is released from some processes within a biorefinery can be used to meet the heat requirements for other lower energy-intensive processes in the system [19].

All these concepts are the basis for understanding what the purpose of a biorefinery is. In some cases, the valorization of residues becomes relevant, first to add value and then to avoid discharge when disposal is not suitable [20]. Another case can be when the main purpose of the biorefinery is the valorization of residues, which is an alternative to an existing efficient disposal method [21]. Biorefineries can also be proposed to improve the global supply chain for different crops such as coffee, palm, or sugarcane (including the maximal use of the residues to obtain high value-added products) [16,22,23]. A specific case can be biorefineries designed for a specific type of products such as energy-producing biorefineries [24].

The authors highlight an open discussion on the biorefinery concept by formulating an open question: Should stand-alone bioenergy-producing facilities be considered as biorefinery systems? Most of the scientific meetings, congresses, and publications include many bioenergy stand-alone processes as a biorefinery topic. Bioenergy/biofuel production systems are indeed of great importance for biorefineries (due to their large demand), but these are just a processing line of many possible products that can be obtained from biomass (however the largest in volume and possibly current interest to decarbonize fossil fuels). In this regard, our proposed definition of biorefinery becomes a complex system, in which biomass is integrally processed or fractionated to obtain more than one product which may include bioenergy (i.e., direct energy), biofuels, chemicals, and high value-added compounds that can only be extracted from bio-based sources. This becomes possible after a comprehensive study of the useful raw materials and a process design based on the latest state-of-the-art technologies and approaches, which include the aspects of the three pillars of sustainability [25].

1.2 General Applications and Conversion Technologies

In this section, we provide a global perspective of biorefinery applications in the real world and technologies of biomass pretreatment, conversion, and separation as the main pillars of these types of processes.

Currently, there are many biorefinery initiatives at different levels of development and implementation. Table 1.1 provides an overview of commercial biorefineries, demonstration plants, pilot plants, and R&D projects that have been built in the United States, Canada, and Europe. The information contained in this table allows us to understand some real applications of these systems, products preferred in industry, process scale, main feedstocks used, and technologies. Some cases presented in Table 1.1 should not be considered as real biorefineries under the current concept discussed previously. Nevertheless, these cases are relevant due to the fact that these projects were designed and implemented in the beginning as plants that can be upgraded into integrated biorefineries. For instance, by using the residues of the same first-generation crop to generate other value-added products or to obtain a particular high marketable product(s) instead of biofuels (one case investigated today is to use the sugar platform for producing chemical building blocks such as lactic acid or furfural in contrast with ethanol) [26].

TABLE 1.1

Biorefineries Developed in the United States, Europe, and Canada [1,10,11,27,28]

Country	Company	Feedstock	Description	Location
United States	Mercurius	Agricultural residue, woody biomass, municipal solid waste (MSW)	Conversion technology: Hybrid (Mercurius's REACH Technology) Primary product: Renewable hydrocarbons (jet fuel and diesel blend stocks) Biofuel capacity (m^3/yr): 95 Scale: Pilot	Ferndale, Washington
	Pacific biogasol	Wheat straw, corn stover and poplar residues	Conversion technology: Biochemical Primary product: Ethanol Co-products: animal feed products (dry corn gluten feed, condensed distillers solubles, corn gluten meal, corn germ, and food-grade yeast) Biofuel capacity (m^3/yr): 10,220 Scale. Demonstration	Boardman, Oregon
	ZeaChem Inc.	Hybrid poplar, stover and cobs	Conversion technology: Thermochemical, gasification Primary product: Ethanol Projected products: lignin, acids, acetates, and alcohols, with expansion opportunities for other biofuels, renewable chemicals, and bio-based products Biofuel capacity (m^3/yr): 946 Scale: Pilot	Boardman, Oregon
	Red Rock Biofuels	Forest residues, corn stover, bagasse, switchgrass, algae	Conversion technology: Thermochemical, gasification Primary product: Renewable hydrocarbon (renewable drop-in jet, diesel, and naphtha fuels) Biofuel capacity (m^3/yr): 60,560 Scale: Pioneer	Lakeview, Oregon

(Continued)

TABLE 1.1 (*Continued*)

Biorefineries Developed in the United States, Europe, and Canada [1,10,11,27,28]

Country	Company	Feedstock	Description	Location
	Amyris Biotechnologies Inc.	Sweet sorghum	Conversion technology: Biochemical Primary product: Renewable hydrocarbon (lubricants, diesel, and jet fuel) Biofuel capacity (m³/yr): 5.2 Scale: Inactive	Emeryville, California
	Logos Technologies	Corn stover, switchgrass, wood chips	Conversion technology: Biochemical Primary product: Ethanol Biofuel capacity (m³/yr): 189.3 Scale: Inactive	Visalia, California
	Fulcrum	MSW	Conversion technology: Thermochemical, gasification Primary product: Renewable hydrocarbons (synthetic crude oil) Biofuel capacity (m³/yr): 39,740 Scale: Pioneer	McCarran, Nevada
	Lignol	Woody biomass (alder, aspen, poplar, and cottonwood)	Conversion technology: Biochemical Primary product: Ethanol Projected products: high-purity lignin and furfural Capacity: 1.8 MGY ethanol, 5,500 tons/yr lignin, 550 tons/yr furfural Scale: Demonstration	Commerce City, Colorado
	Rentech Clear Fuels Technology	Woody waste and bagasse	Conversion technology: Thermochemical, gasification Primary product: Renewable hydrocarbons (hydrogen and syngas) Biofuel capacity (m³/yr): 571.6, 151,000 Scale: Inactive	Commerce City, Colorado

(*Continued*)

TABLE 1.1 (*Continued*)

Biorefineries Developed in the United States, Europe, and Canada [1,10,11,27,28]

Country	Company	Feedstock	Description	Location
	Abengoa Bioenergy Biomass of Kansas (ABBK)	Stover, switchgrass, woody biomass	Conversion technology: Biochemical Primary product: Ethanol Co-product: Electricity (300 tons/day of raw biomass material to produce 18 MW) Biofuel capacity (m³/yr): 94,630 Scale: Pioneer	Hugoton, Kansas
	Sapphire Energy Inc.	Algae	Conversion technology: Algae Primary product: Renewable hydrocarbon (fuels) Co-products: Omega-3 oils, high-value aquaculture, animal feed ingredients Biofuel capacity (m³/yr): 3,785 Scale: Demonstration	Columbus, New Mexico
	Verenium	Sugarcane bagasse, energy cane, and sorghum	Conversion technology: Biochemical Primary product: Ethanol Biofuel capacity (m³/yr): 5,300 Scale: Demonstration (inactive)	Jennings, Louisiana
	Emerald Biofuels	Inedible corn oil, food processing waste, animal fats, and yellow/brown greases	Conversion technology: Thermochemical—hydrogenated esters and fatty acids Primary product: Renewable carbons (diesel) Biofuel capacity (m³/yr): 310,400 Scale: Pioneer	Plaquemine, Louisiana
	Myriant Succinic Acid Biorefinery (MySAB)	Sorghum, multi-feedstock	Conversion technology: Biochemical Primary product: Bio-succinic acid Capacity: 30 million pounds per year Scale: Demonstration	Port of Lake Providence, Louisiana

(*Continued*)

TABLE 1.1 (*Continued*)

Biorefineries Developed in the United States, Europe, and Canada [1,10,11,27,28]

Country	Company	Feedstock	Description	Location
	Algenol Biofuels Inc.	Carbon dioxide, algae, seawater	Conversion technology: Algae Primary product: Ethanol Biofuel capacity (m^3/yr): 378.5 Scale: Pilot	Fort Meyers, Florida
	INEOS New Planet Bioenergy LLC	Vegetative	Conversion technology: Thermochemical, gasification Primary product: Ethanol Co-product: Electricity Biofuel capacity (m^3/yr): 30,280 Scale: Pioneer	Vero Beach, Florida
	Bioprocess Algae	Algae	Conversion technology: Algae Primary product: Renewable hydrocarbons Scale: Pilot	Shenandoah, Iowa
	ICM Inc.	Corn fiber, switchgrass, energy sorghum	Conversion technology: Biochemical Primary product: Ethanol Biofuel capacity (m^3/yr): 927 Scale: Pilot	St. Joseph, Missouri
	POET Project LIBERTY, LLC	Corn cobs	Conversion technology: Biochemical Primary product: Ethanol Co-product: Methane gas Biofuel capacity (m^3/yr): 75,700 Scale: Pioneer	Emmetsburg, Iowa
	Frontline bioenergy LLC	Woody biomass and MSW	Conversion technology: Thermochemical, gasification Primary product: Renewable hydrocarbons (syngas as platform for the production of diesel, avgas, jet fuel, methanol, ethanol, biobutanol) Biofuel capacity (m^3/yr): 75.7 Scale: Pilot	Ames, Iowa

(*Continued*)

TABLE 1.1 (*Continued*)

Biorefineries Developed in the United States, Europe, and Canada [1,10,11,27,28]

Country	Company	Feedstock	Description	Location
	Flambeau River Biofuels LLC	Wood and forest residues	Conversion technology: Thermochemical, gasification Primary product: Renewable hydrocarbons (green diesel) Biofuel capacity (m³/yr): 34,000 Scale: Demonstration	Park Falls, Wisconsin
	NewPage	Mill residues and un-merchantable wood	Conversion technology: Thermochemical, gasification Primary product: Fischer–Tropsch (F–T) liquids (renewable diesel) and heat (offset natural gas) Capacity (m³/yr): 31,040 of F–T liquids Scale: Demonstration	Wisconsin Rapids, Wisconsin
	Solazyme Inc.	Algae	Conversion technology: Algae Primary product: Renewable hydrocarbons (biodiesel and renewable diesel from purified algal oil) Capacity (m³/yr): 1,135 of purified algal oil Scale: Pilot	Peoria, Illinois
	Archer Daniels Midlan—ADM	Corn stover	Conversion technology: Biochemical Primary product: Ethanol Biofuel capacity (m³/yr): 98 Scale: Pilot	Decatur, Illinois
	HaldorTopsoe Inc.	Wood waste and nonmerchantable wood	Conversion technology: Thermochemical, gasification Primary product: Renewable hydrocarbons (gasoline) Biofuel capacity (m³/yr): 1,306 Scale: Pilot	Des Plaines, Illinois
	Elevance Renewable Sciences	Natural oils such as algal oil, and other seed oils and animal fats	Conversion technology: Chemical Primary product: Renewable hydrocarbons (green diesel, biodiesel, alpha olefins, methyl esters for surfactant) Scale: Design	Bolingbrook, Illinois

(*Continued*)

TABLE 1.1 (*Continued*)

Biorefineries Developed in the United States, Europe, and Canada [1,10,11,27,28]

Country	Company	Feedstock	Description	Location
	Gas Technology Institute—GTI	Wood waste, corn stover and algae	Conversion technology: Thermochemical, pyrolysis Primary product: Renewable hydrocarbons Scale: Design	Des Plaines, Illinois
	Mascoma	Hardwood pulpwood	Conversion technology: Biochemical Primary product: Ethanol Co-products: Lignin and bark used to produce heat and electricity Biofuel capacity (m^3/yr): 75,700 Scale: Pioneer	Kinross Charter township, Michigan
	American Process Inc.—API. Alpena Biorefinery	Mixed northern hardwood and aspen	Conversion technology: Biochemical Primary product: Ethanol and potassium acetate Capacity (m^3/yr): 3,380 of cellulosic ethanol and 2,630 of aqueous potassium acetate Scale: Pilot	Alpena, Michigan
	Renewable Energy Institute—REII	Rice hulls and forest residues	Conversion technology: Thermochemical, gasification Primary product: Renewable hydrocarbons Biofuel capacity (m^3/yr): 2,360 Scale: Pilot	Toledo, Ohio
	Red Shield Acquisition (RSA), LLC	Forest resources	Conversion technology: Biochemical Primary product: Lignocellulosic sugars, algal oil, and ethanol Capacity: 28,000 tons per year of sugars, 2,100 m^3 per year of green oil, and 49,200 m^3 per year of ethanol Scale: Demonstration	Old town, Maine
	UOP, LLC	Agricultural and forestry residue, wood, energy crops, and algae	Conversion technology: Thermochemical—pyrolysis Primary product: Renewable hydrocarbons (gasoline, diesel, and jet fuels) Biofuel capacity (m^3/yr): 227 Scale: Pilot	Kapolei, Oahu, Hawaii

(*Continued*)

TABLE 1.1 (*Continued*)

Biorefineries Developed in the United States, Europe, and Canada [1,10,11,27,28]

Country	Company	Feedstock	Description	Location
Austria	Green biorefinery (Bioraffinerie Forschung und Entwicklung GmbH)	Mixtures of grass, clover, lucerne silage	Conversion technology: Mechanical fractionation Platform: Biogas and organic solution Products: Organic acids, biomaterials, fertilizer, biomethane, or electricity and heat Scale: Demonstration	Linz
	M-real, Hallein AG	Sprucewood	Conversion technology: Fermentation Platform: C6 sugars Products: Bioethanol, biomaterials, electricity and heat Scale: Conceptual	Hallein
	Lenzing (Lenzing Pulp Mill)	Fiber and pulp	Conversion technology: Cooking, extraction. Platform: C5 sugars, lignin, electricity and heat Products: Furfural, cellulosic fibers, food-grade acetic acid, xylose-based artificial sweetener Capacity: 470,000 tons of acetic acid and about 110,000 tons of furfural Scale: Commercial	Lenzing
	AGRANA Bioethanol GmbH (Pischelsdorf Biorefinery)	Cereals	Platform: Starch, C5/C6 sugars Products: Starch and gluten, bioethanol, animal feed, CO_2 Capacity (m^3/yr): 240,000 ethanol Scale: Commercial	Pischelsdorf
	Zellstoff Pöls AG (Pöls Biorefinery)	Wood	Conversion technology: Cooking, separation Platform: Pulp, black liquor, and electricity and heat Products: Pulp, paper, tall oil, turpentine, bark, electricity and heat Scale: Commercial	Pöls

(*Continued*)

TABLE 1.1 (*Continued*)

12

Biorefineries Developed in the United States, Europe, and Canada [1,10,11,27,28]

Country	Company	Feedstock	Description	Location
Belgium	Tereos Starch & Sweeteners Europe (Aalst Wheat Processing Plant)	Wheat	Conversion technology: Hydrolysis, fermentation Platform: Starch, C5/C6 sugars Products: Refined glucose syrups, bioethanol Scale: Commercial	Aalst
	BioWanze (Wanze Bioethanol Plant)	Wheat, sugar syrups	Conversion technology: Hydrolysis, fermentation Platform: Starch, C5/C6 sugars Products: Bioethanol, gluten, soluble protein concentrate Capacity (m^3/yr): 300,000 ethanol Scale: Commercial	Wanze
Denmark	Inbicon IBUS	Straw	Conversion technology: Hydrolysis, fermentation Platform: C6/C5 sugars and lignin Products: Bioethanol, animal feed (molasses with hemicellulose sugars), lignin, electricity and heat Scale: Pilot	Fredericia
Finland	MetsäFibre (MetsäBioproduct Mill)	Wood	Conversion technology: Cooking, separation Platform: Pulp, lignin, electricity and heat Products: Pulp, electricity, tall oil, turpentine, lignin products and wood fuel. Potential new products: producer gas, sulfuric acid, textile fibers, biocomposites, fertilizers, and biogas. Scale: Under construction	Äänekoski
	Fortum (Joensuu Bio-oil Plant)	Forest residues and other wood-based biomass	Conversion technology: Pyrolysis Platform: Pyrolysis oil and heat and electricity Products: Bio-oil and heat and electricity Capacity (tons/yr): 50,000 bio-oil Scale: Commercial	Joensuu

(Continued)

TABLE 1.1 (*Continued*)

Biorefineries Developed in the United States, Europe, and Canada [1,10,11,27,28]

Country	Company	Feedstock	Description	Location
France	UPM Biofuels (UPM Lappeeranta Biorefinery)	Tall oil	Conversion technology: Pretreatment, hydrotreatment Platform: oil Products: Biodiesel Scale: Commercial	Lappeenranta
	Pomacle-Bazancourt Biorefinery	Sugar beet, wheat	Conversion technology: Hydrolysis, fermentation, extraction Platform: Starch, C5/C6 sugars Products: Starch, glucose, animal feed, bioethanol, CO_2 Scale: Commercial (although it also covers pilot plants and labs)	Pomacle-Bazancourt
	Roquette (Lestrem Starch Biorefinery)	Wheat, potato, maize, pea	Conversion technology: Hydrolysis, fermentation Platform: Starch, C5/C6 sugars Products: Starch, food, feed, bulk and fine chemicals, succinic acid, ethanol Scale: Commercial	Lestrem
	Sofiproteol	Rape, sunflower oilseed	Conversion technology: Pressing, esterification Platform: Starch, oil Products: Biodiesel, glycerin, chemicals and polymers (coatings, biolubricants, biopolymers), animal feed Scale: Commercial	N.A.
Germany	CHOREN (Shell Deutschland Oil GmbH, Volkswagen AG, Daimler AG)	Woody biomass	Conversion technology: Gasification, combustion Platform: Syngas Products: Synthetic biofuels or electricity and heat Scale: Demonstration	Freiberg
	Zellstoff Stendal GmbH (Mercer International Inc)	Softwood, small logs, sawmill chips	Conversion technology: Combustion Platform: Lignin Products: Biomaterials, electricity and heat Scale: Commercial	Arneburg

(Continued)

TABLE 1.1 (*Continued*)

Biorefineries Developed in the United States, Europe, and Canada [1,10,11,27,28]

Country	Company	Feedstock	Description	Location
	Biowert GmbH (Biowert Biorefinery)	Grass	Conversion technology: Pressing, drying, separation Platform: Organic solutions (green juice and fibers), biogas, electricity and heat Products: High-quality cellulosic fibers, nutrients, biogas Scale: Commercial	Brensbach
	Cargill (Krefeld Starch Biorefinery)	Maize	Conversion technology: Hydrolysis, separation Platform: Starch Products: Starch, dextrose, sorbitol, glucose, isomalts, maltitol syrup, feed Scale: Commercial	Krefeld
	Verbio (Schwedt Biorefinery)	Rye	Conversion technology: Hydrolysis, fermentation Platform: Starch, C5/C6 sugars Products: Bioethanol, biogas, organic fertilizers Capacity (tons/yr): 180,000 ethanol Scale: Commercial	Schwedt
	Crop Energies Bioethanol GmbH	Sugar beet, grain	Conversion technology: Hydrolysis, fermentation Platform: Starch, C5/C6 sugars Products: Bioethanol, animal feed, CO_2 Capacity (m³/yr): 360,000 ethanol Scale: Commercial	Zeitz
	Verbio (Zörbig Biorefinery)	Rye, triticale, wheat	Conversion technology: Hydrolysis, fermentation Platform: Starch, C5/C6 sugars Products: Bioethanol, biogas, organic fertilizers Capacity (tons/yr): 90,000 ethanol Scale: Commercial	Zörbig

(Continued)

TABLE 1.1 (*Continued*)

Biorefineries Developed in the United States, Europe, and Canada [1,10,11,27,28]

Country	Company	Feedstock	Description	Location
Hungary	Pannonia Ethanol (Dunaföldvár Biorefinery)	Corn	Conversion technology: Hydrolysis, fermentation Platform. Starch, C5/C6 sugars Products: Bioethanol, animal feed and corn oil Capacity (L/yr): 450 million ethanol Scale: Commercial	Dunaföldvár
Italy	GFBiochemicals (Caserta Levulinic Acid Plant)	Cellulosic biomass	Conversion technology: Acid hydrolysis Platform: C5/C6 sugars Products: Levulinic acid Scale: Commercial	Caserta
	Reverdia (Cassano Succinic Acid Plant)	Starch and sugars	Conversion technology: Hydrolysis, fermentation Platform: C5/C6 sugars Products: Succinic acid Scale: Commercial	CassanoSpinola
	Beta Renewables (Crescentino Bioethanol Plant)	Giant reed, miscanthus, switch grass, agricultural waste (straws)	Conversion technology: Hydrolysis, fermentation Platform: C5/C6 sugars, lignin, power and heat Products: Bioethanol and animal feed Capacity (tons/yr): 40,000 ethanol Scale: Commercial	Crescentino
	Novamont (Mater-Biotech Plant)	Sugar (glucose syrup produced by Cargill)	Conversion technology: Fermentation Platform: C5/C6 sugars, biogas Products: 1,4-butanediol (BDO) Capacity (tons/yr): 30,000 BDO Scale: Commercial	Bottrighe
	Novamont and Versalis (Matrica Biorefinery)	Oil-seed from nonfood autochthonous crops	Platform: Oil Products: Monomers, rubber additives, biolubrificants, bioplastics Capacity (tons/yr): 35,000 Scale: Commercial	Matrica

(*Continued*)

TABLE 1.1 (*Continued*)

Biorefineries Developed in the United States, Europe, and Canada [1,10,11,27,28]

Country	Company	Feedstock	Description	Location
	eni (Venice Biorefinery)	Vegetable oils, animal fats, used cooking oils	Conversion technology: Hydrotreatment Platform: Oil Products: Hydrotreated vegetable oil (HVO) Capacity (tons/yr): 300,000 green diesel Scale: Commercial	Porto Marghera, Venice
Netherlands	European Bio-Hub Rotterdam	Lignocellulosic biomass	Platform. Syngas, C5/C6 sugars Products: Base chemicals, biofuels Scale: Demonstration	Rotterdam
	BioMCN (DelfzijlBiomethanol Plant)	Glycerin	Conversion technology: Evaporation, reforming Platform: Syngas Products: Biomethanol Scale: Commercial	Delfzijl
	WUR Micro-algae Biorefinery	Microalgae	Platform: Oil and organic solution Products: base chemicals, bio-fuels Scale: Pilot	Wageningen
	Biodiesel Amsterdam, Noba, Orgaworld and Rotie (Green mills Project)	Fryer fat, other organic waste	Conversion technology: Esterification, fermentation Platform: Starch, C5/C6 sugars Products: Biodiesel, bioethanol, biogas, fertilizers Capacity (yr): 25 million m^3 biogas, 113 million liters of biodiesel, and 5 million liters of bioethanol Scale: Commercial	Port of Amsterdam
	Neste Oil (Rotterdam Biorefinery)	Vegetable oil, waste animal fat	Conversion technology: Hydrotreatment Platform: Oil Products: HVO Capacity (tons/yr): 1 million Scale: Commercial	Rotterdam

(Continued)

TABLE 1.1 (*Continued*)

Biorefineries Developed in the United States, Europe, and Canada [1,10,11,27,28]

Country	Company	Feedstock	Description	Location
	Cargill (Sas van Gent Biorefinery)	Wheat and corn at Cargill; wheat by-products of Cargill plant to Nedalco	Conversion technology: Hydrolysis, fermentation Platform: Starch, C5/C6 sugars Products: Starches, starch derivatives, wheat proteins and glucose, bioethanol Scale: Commercial	Sas van Gent
Norway	Borregaard (SarpsborgBiorefinery)	Lignocellulosic crops or residues	Conversion technology: Cooking, fermentation, separation Platform: C5/C6 sugars, lignin, biogas Products: Bioethanol, lignin, cellulose, vanillin, biogas Scale: Commercial	Sarpsborg
Spain	Succinity GmbH (Montmeló Succinic Acid Plant)	Glycerol, sugars	Conversion technology: Hydrolysis, fermentation Platform: C5/C6 sugars Products: Succinic acid Capacity (tons/yr): 10,000 Scale: Commercial	Montmeló
Sweden	Aditya Birla Group (Domsjö Pulp Mill)	Lignocellulosic crops	Conversion technology: Cooking, fermetantion, separation Platform: C5/C6 sugars, lignin, biogas Products: Bioethanol, lignin, cellulose Scale: Commercial	Domsjö
	SunPine (Piteå Tall-Oil Biorefinery)	Crude tall oil	Conversion technology: Esterification, extraction Platform: Oil Products: Crude tall diesel, bio-oil, rosin Scale: Commercial	

(Continued)

TABLE 1.1 (*Continued*)

Biorefineries Developed in the United States, Europe, and Canada [1,10,11,27,28]

Country	Company	Feedstock	Description	Location
United Kingdom	Ensus (Ensus Wheat Biorefinery)	Wheat	Conversion technology: Hydrolysis, fermentation Platform: C5/C6 sugars Products: Bioethanol, dried distillers grains with solubles (DDGS), CO_2 Capacity (yr): 400,000 cubic meters of bioethanol and 350,000 tons of dried protein animal feed (DDGS) Scale: Commercial	Teesside
	Vivergo Fuels (Hull Biorefinery)	Starch crops, feed grade wheat	Conversion technology: Hydrolysis, fermentation Platform: C6 sugars Products: Bioethanol, animal feed Capacity (m^3/yr): 420,000 ethanol Scale: Commercial	Saltend Chemicals Park, Hull
	Cargill (Manchester Biorefinery)	Wheat and corn at Cargill, wheat by-products of Cargill plant to Nedalco	Conversion technology: Hydrolysis, fermentation Platform: Starch, C5/C6 sugars Products: Starches, starch derivatives, wheat proteins and glucose, bioethanol Scale: Commercial	Manchester
	British Sugar (Wissington Factory)	Sugar beet	Conversion technology: Fermentation, anaerobic digestion Platform: C5/C6 sugars, power/heat Products: Sugar products, bioethanol, animal feed, lime, tomatoes, topsoil, power and heat Capacity (tons/yr): 55,000 ethanol Scale: Commercial	Wissington
Canada	Permolex International, L.P	Wheat	Conversion technology: Mechanical fractionation, enzymatic hydrolysis, fermentation Platform: C6 sugar Products: bioethanol, animal feed, food (bakery flour, gluten) Scale: Commercial	Red Deer, Alberta

(Continued)

TABLE 1.1 (*Continued*)

Biorefineries Developed in the United States, Europe, and Canada [1,10,11,27,28]

Country	Company	Feedstock	Description	Location
	Lignol (Lignol Innovations Ltd.)	Softwood, hardwood, annual fibers, agro-residuals	Conversion technology: Hydrolysis, fermentation Platform: C6/C5 sugars and lignin Products: Bioethanol, chemicals (furfural), biomaterials Scale: Pilot	Burnaby, BC
	Ensyn	Residual woody biomass from a hardwood flooring plant and sawmill	Conversion technology: Pyrolysis Products: food (flavorings), polymers (resins), heat, electricity plus (in the future) synthetic biofuels Scale: Commercial	Ontario
	Highmark Renewables	Wheat, manure, slaughtering waste	Conversion technology: Hydrolysis, fermentation, combustion Platform: C6 sugars, biogas Products: Bioethanol, animal feed, fertilizer, electricity and heat Scale: Commercial	Vegreville, Alberta

1.2.1 Pretreatment Methods of Biomass

The goal of a conversion process is to transform the feedstock into end products. Some renewable feedstocks such as lignocellulose need to be conditioned and therefore need to be submitted to a pretreatment stage to release the sugars contained within the matrix of material. The pretreatment process is a crucial stage of biomass conversion into value-added end products. This first stage usually will define the quality and success of the desired biorefinery based on the integral use of the biomass. An ideal pretreatment has the following goals: (i) to recover the major possible quantity of carbohydrates; (ii) to guarantee a high digestibility of cellulose in the subsequent step, enzymatic hydrolysis; (iii) to present null or low amounts of degraded sugars and products from lignin; (iv) to avoid the formation of inhibitors; (v) to improve the concentration of the released sugars in liquid fraction; and (vi) low energy consumption (vii) to be economically feasible [29–32].

In general terms, the pretreatments are divided in groups such as physical (e.g., chipping, grinding, milling, irradiation by γ-rays, ultrasonic and pyrolysis), chemical (e.g., ozonolysis, acid hydrolysis, alkaline hydrolysis, wet oxidation, oxidative delignification, organosolv, and ionic liquids), biological (e.g., fungal and bio-organosolv), and combinations between them such as physicochemical (e.g., steam explosion, liquid hot water (LHW) or hydrothermolysis, ammonia fiber explosion (AFEX), and CO_2 explosion) [29–31,33–35]. Table 1.2 presents the classification of pretreatment methods of biomass and a brief description of each one.

The physical methods can be combined with chemical methods to have better access to the material by part of reagents considered and, therefore, better yields to reducing sugars (i.e., pentoses and hexoses). Physical pretreatments can increase the available surface area and reduce both the cellulose crystallinity and the degree of polymerization [33]. The energy consumption in physical pretreatments depends on the particle size required and biomass characterization (e.g., agricultural waste requires less energy than hardwood) [30,33].

1.2.2 Conversion and Separation Methods of Biomass

The methods to separate, refine, and transform biomass in energy, chemicals, and biofuels can be categorized into five groups: physical, thermochemical, biological, chemical, and hybrid [1,37–39]. Table 1.3 provides a brief description for each conversion/separation technique and some examples.

1.3 Current Design Approaches

The design approach for biorefineries is still an important aspect to discuss due to the difficulty and the infinite possibilities of portfolios to assess. Due

TABLE 1.2

Pretreatment Methods of Lignocellulosic Materials [29–31,33,34,36]

Method	Technique	Remarks
Physical	Chipping	Reduces the biomass size to 10–30 mm. The goal of this technique is to reduce heat and mass transfer limitations.
	Grinding Milling	Reduces the biomass size to 0.2–2 mm. These techniques are more effective in reducing the cellulose crystallinity and particle size.
	Irradiation by γ-rays	Cleaves the β-1,4-glycosidic bonds of material. This technique is very expensive at high scale.
	Ultrasonic	The ultrasonification pretreatment has been studied at a laboratory scale.
	Pyrolysis	In pyrolysis pretreatment, the cellulose rapidly decomposes at temperatures greater than 300°C in gaseous products and residual char. Residues can undergo mild dilute acid hydrolysis (1N H_2SO_4, 2.5 h, 97°C) to produce 80%–85% reducing sugars (>50% glucose). It can be carried out under vacuum (400°C, 1 mmHg, 20 min).
Chemical	Ozonolysis	The ozone degrades the lignin and hemicellulose in lignocellulosic biomass. The ozonolysis reactions are carried out at room temperature and atmospheric pressure. There is no formation of inhibitors. The ozonolysis can require a large amount of ozone making the pretreatment expensive.
	Acid hydrolysis	*Dilute acid.* 0.75%–5% H_2SO_4, HCl, and HNO_3 can be used as reagents at 1 MPa. Continuous process for low solids loads (5–10 wt.% dry substrate/mixture): 160°C–200°C. Batch process for high solids loads (10–40 wt.% dry substrate/mixture): 120°C–160°C. *Concentrated acid.* 10%–30% H_2SO_4, 170°C–190°C, 1:1.6 solid–liquid ratio. The acid hydrolysis pretreatment improves the enzymatic hydrolysis yields to reduce sugars. The acids used in this pretreatment are toxic and corrosive, and therefore need equipment of resistant materials making the process slightly expensive. This pretreatment can be used in a wide range of raw materials from hardwoods to agricultural wastes. Inhibitor formation is present. Acid pretreatment needs alkali to neutralize the hydrolyzate.

(Continued)

TABLE 1.2 (*Continued*)

Pretreatment Methods of Lignocellulosic Materials [29–31,33,34,36]

Method	Technique	Remarks
	Alkaline hydrolysis	Dilute NaOH, 24 h, 60°C; Ca(OH)$_2$, 4 h, 120°C. NaOH, KOH, Ca(OH)$_2$, hydrazine and anhydrous ammonia can be used as reagents. Alkaline hydrolysis is more effective in biomass with low lignin content. Formation of inhibitor is less. This pretreatment can be developed at room conditions but requiring large number of hours.
	Oxidative delignification	Peroxidase and 2% H$_2$O$_2$, 20°C, 8 h. Almost total solubilization of hemicellulose and 50% of the lignin. Cellulose conversion in enzymatic hydrolysis can be 95%.
	Wet oxidation	1.2 MPa oxygen pressure, 195°C, 15 min; addition of water and small amounts of Na$_2$CO$_3$ or H$_2$SO$_4$. Wet oxidation pretreatment is presented as alternative to steam explosion. This pretreatment uses high temperatures (150–350°C) and high pressure (5–20 MPa). This technique can be used for treatment of agricultural residues and hardwood. Formation of inhibitor is present. It is efficient for pretreatment of lignocellulosic biomass due to the crystalline structure of cellulose that is exposed.
	Organosolv process	Organic solvents (methanol, ethanol, acetone, ethylene glycol, triethylene glycol, and tetrahydrofurfuryl alcohol) or their mixture with 1% of H$_2$SO$_4$ or HCl; 185°C–198°C, 30–60 min, pH=2.0–3.4. This pretreatment breaks internal hemicellulose and lignin bonds. Almost total solubilization of lignin. Solvent recovery required in order to reduce the cost and avoid inhibition in the next reactions such as enzymatic hydrolysis and fermentation.
	Ionic liquids	The pretreatment with ionic liquids uses components that due to their unique properties and polarity can act as selective solvents to cellulose or lignin. This technique could avoid the inhibitor formation and operate at ambient conditions.

(Continued)

TABLE 1.2 (*Continued*)

Pretreatment Methods of Lignocellulosic Materials [29–31,33,34,36]

Method	Technique	Remarks
Biological	Fungal pretreatment	Brown-, white- and soft-rot fungi are used for hemicellulose and lignin degrading. The pretreatment is environmentally friendly and demands low amount of energy, but the rate of biological hydrolysis is very low for industrial use and requires long residence times. The technique presents loss of material because the fungi consumes the biomass.
	Bioorganosolv pretreatment	*Ceriporiopsis subvermispora* for 2–8 weeks followed by ethanolysis at 140°C–200°C for 2h.
Physicochemical	Steam explosion	The biomass is treated with saturated steam at 160°C–290°C, 0.69–4.85MPa for several seconds or minutes, then decompression until atmospheric pressure. This technique is widely used, and the hemicellulose is the component predominantly solubilized and found in the liquid phase as sugars. Steam pretreatment can be improved with an acid catalyst such as H_2SO_4 or SO_2 increasing the recovery of hemicellulosic sugars. This pretreatment has been applied at a pilot and a demonstration scale.
	LHW or hydrothermolysis	Pressurized hot water, $P > 5$ MPa, 170°C–230°C, 1–46 min; solids load <20%. This pretreatment is similar to steam explosion with lower temperatures and requires great amount of water, demanding energy in the downstream process.
	AFEX	1–2 kg ammonia/kg dry biomass, 90°C, 30 min, 1.12–1.36 MPa. In this technique, large amount of the hemicellulose is degraded to sugars and small amount of solid is solubilized. AFEX presents good performance with agricultural residues and is inefficient with wood because it presents high lignin content. There is no formation of inhibitors that can affect the next biological reactions.
	CO_2 explosion	4 kg CO_2/kg fiber, 5.62 MPa. CO_2 explosion operates at low temperatures avoiding the sugars degradation. There is no formation of inhibitors.

TABLE 1.3

Classification of Methods to Transform and Separate Biomass [1, 37–39]

Method	Technique	Remarks	Example
Physical	Distillation	Separates the interesting component from a liquid mixture by selective evaporation and condensation.	Ethanol and butanol purification
	Supercritical fluid	Uses supercritical fluids (e.g., CO_2) to separate one component (extract) from the other (solid matrix). It can also be used for extractions from liquids.	Phenolic compounds, oils
	Solvent extraction	Uses solvents (e.g., water, ethanol, hexane, toluene, isopropyl alcohol, etc.) to separate one component (extract) from the other (solid matrix) at reasonable temperatures (20°C–90°C) and atmospheric pressure.	Phenolic compounds and oils
	Crystallization	Forms crystals through precipitation from a solution, melting, or deposition directly from a gas. The characteristics of the crystal depend on the temperature, time of fluid evaporation, or air pressure.	Xylitol and sorbitol
Thermochemical	Combustion	Consists in the thermal conversion (burn) of biomass in the presence of an oxidant (normally O_2/air)	Heat, CO_2, and H_2O
	Gasification	Consists in the thermal decomposition of biomass at high temperatures (800°C–1300°C) with limited O_2 levels.	Syngas (i.e., H_2, CO, CO_2, and CH_4). The gas can be used for the production of energy by means of turbines or gas engines.
	Hydrothermal upgrading	Uses supercritical water (300°C–350°C, 10–18 MPa) during 5–20 min to generate a heavy organic liquid ("biocrude"). In this process, there is a reduction of oxygen in the organic material from 40% to 10%–15%.	Cellobiose, glucose, and hydroxymethylfurfural (HMF)

(Continued)

TABLE 1.3 (*Continued*)

Classification of Methods to Transform and Separate Biomass [1, 37–39]

Method	Technique	Remarks	Example
	Pyrolysis	Involves heating the biomass to intermediate temperatures (300°C–600°C) in the absence of oxygen, to break the structure of the present polymers and convert them into a liquid product, a gas rich in hydrocarbons, and a solid residue	Liquid pyrolytic oil (i.e., bio-oil), charcoal, and light gases
	Liquefaction	Hydrolyzes or degrades the biomass macromolecules by means of water (supercritical) at average temperatures (280°C–370°C) and high pressures (10–25 MPa).	Bio-oil
Biological	Fermentation	Uses microorganisms that under certain nutritional conditions and at different conditions of oxygen can transform a substrate in extracellular or intracellular biomolecules at low temperature and low reaction rate.	Biofuels (e.g., ethanol, butanol, biogas, hydrogen, etc.), drugs (e.g., penicillin), food (e.g., beer, yogurt, bread), and biopolymers (e.g., polyhydroxybutyric acid (PHB) and polyols)
	Anaerobic digestion	Uses bacteria for the decomposition of biomass in the absence of oxygen at 30°C–65°C.	Biogas (i.e., CH_4 and CO_2), hydrogen
	Aerobic digestion	Uses bacteria for the decomposition of biomass in the presence of oxygen.	Sewage treatment, antibiotics, some biopolymers, terpenes, organic acids, and amino acids
	Enzymatic hydrolysis	Uses enzymes to transform a substrate in recoverable products. Enzyme charge is based on the enzymatic activity (U), 1:10 solid–liquid, pH= 4.5–5.5, 50°C–60°C	C6 sugars (i.e., glucose)
Chemical	Esterification/ transesterification	Reaction between vegetable oils and ethanol or methanol in the presence of KOH or NaOH as a catalyst to generate methyl or ethyl esters of fatty acids.	Biodiesel

(*Continued*)

TABLE 1.3 (*Continued*)

Classification of Methods to Transform and Separate Biomass [1, 37–39]

Method	Technique	Remarks	Example
	Hydrogenation	Reaction between H_2 and other compounds to produce biomolecules, usually in the presence of a catalyst such as platinum, palladium, or nickel. The CO and CO_2 hydrogenation and the reverse water–gas shift reactions are performed to produce methanol over a Cu-based catalyst.	Furfural and levulinic acid (bio-derived molecules) can be hydrogenated to obtain furfuryl alcohol and γ-valerolactone, respectively, over a noble metal-free Cu–Cr catalyst
	Dehydration	Removes water molecules from sugars, fructose, glucose, or xylose to produce chemical building blocks in the presence of an acid catalyst.	Furfural and HMF
	Other examples: methanation, steam reforming, water electrolysis, water–gas shift, oxidation, etc.		
Hybrid	Simultaneous saccharification and fermentation	Combines enzymatic hydrolysis with fermentation in a single step.	Biofuels (i.e., ethanol and butanol)
	Simultaneous saccharification and co-fermentation	Uses an enzymatic complex to decompose the cellulose in sugars such as hexoses. These sugars produced *in situ* together with pentoses after pretreatment are consumed simultaneously by specialized microorganisms that have the ability to consume the two substrates.	Ethanol

to aspects such as, raw materials, technologies, products, context, conversion routes, social aspects, logistics, policies, laws and many other elements, the biorefineries turn into an open and complex problem. This shows the necessity of including a deep view depending on the context and the impacted region. This allows developing patterns for local solutions. Therefore, the different design approaches depend on the restrictions for systems, adjacent markets, and especially the population to be impacted [8]. For these reasons, it is essential to implement a methodology, method, or model to choose the optimal routes and the best products. The three main approaches for the design of biorefineries developed until now are described in Sections 1.3.1–1.3.3.

1.3.1 Conceptual Design

The conceptual design promotes an approach based on an inclusive design of biorefineries, where new challenges are assumed such as the inclusion of a wide range of feedstocks and the development of local and regional areas. As this methodology includes many factors, it is possible to say that it is based on a holistic approach. This strategy has been applied in the process engineering for the design of chemical plants but not for the design of biorefineries [40,41]. Characteristics such as availability, costs, and market needs are the basis to choose the feedstocks and products. However, there are some constraints linked to uses and applications of the main streams. For example, the value-added products such as antioxidants have priority in their production over energy products such as biofuels.

Three types of analysis are included in the conceptual design of a biorefinery: (i) technical, (ii) economic, and (iii) environmental. The commercial software package Aspen Plus and SuperPro Designer are the main tools used to generate mass and energy balances related to the technologies. Then, Aspen Process Economic Analyzer is used to calculate the economic results. The U.S. Environmental Protection Agency (EPA) Waste Reduction (WAR) algorithm and/or GREENSCOPE allow estimating the generated environmental impact [42–44]. Two types of integration are then applied: (i) mass using process modeling and (ii) energy using pinch or energy analysis. Experimental data such as feedstock characterization, reaction kinetics, yields, and operation conditions reported in literature are fed in the simulation tool. A comparison between the reported data is made in order to choose those presenting the best performance. The conceptual design applied to biorefineries comes from the approach of synthesis of chemical processes [40,41]. However, the approach is extended to complex systems (i.e., biorefineries with a multiproduct portfolio) because new elements are included. Three concepts compose the conceptual design: (i) hierarchy, (ii) sequencing, and (iii) integration [8,40,45]. These concepts are interlinked, and their combination leads to a process synthesis approach to obtain biorefinery models. A brief description of these concepts is presented as follows:

1. *Hierarchy.* This concept considers the hierarchical decomposition of the main elements in a biorefinery such as feedstocks, products, and technologies. The first analysis is focused on feedstocks and products, where the feedstock composition and possible decomposition in platforms to generate a set of products are taken into account. The second analysis is focused on the technologies and their critic points. An extension of the well-known onion diagram is considered to design the process schemes and processing routes that compose a biorefinery [9,18].

2. *Sequencing.* This concept establishes a logical order step-by-step to associate transformation routes and products. The implementation of

this concept considers some factors such as (i) platforms, (ii) development in the technological lines, and (iii) constraints in the quality of product. The determination of the most promising sequence requires systematic tools to identify the routes, distributions, platforms, and products possessing the best characteristics. Optimization is a clear tool used in the decision-making [9,18].

3. *Integration.* This concept defines the possibilities of integration between feedstocks, processing routes, and products. It emphasizes the maximum use of resources in the biorefinery. In the case of feedstocks, the remaining streams of a process can be exploited in other processes in order to reduce the waste streams. For processing routes, there are two types of integration: (i) mass and (ii) energy that are directly linked to economic and environmental aspects [8,40,45].

1.3.2 Optimization

Optimization can assist the bioprocessing industries in deciding which routes have the maximal profitability of transformation and product portfolios; meanwhile, the stakeholder value is maximized through the global optimization of supply chain. Many authors have applied the optimization methodology to the design of all type of biorefineries, for example, Sammons Jr et al. (2007) [46], Sammons Jr et al. (2008) [47], Zondervan et al. (2011) [48], Bertran et al. (2017) [49], Santibañez-Aguilar et al. (2011) [50], Giarola et al. (2011) [51], Ponce-Ortega et al. (2012) [52], Pham & El-Halwagi (2012) [53], You et al. (2012) [54], El-Halwagi et al. (2013) [55], Wang et al. (2013) [56], Tong et al. (2014) [57], Kelloway et al. (2014) [58], Gong et al. (2014) [59], Santibañez-Aguilar et al. (2014) [60], Geraili et al. (2015) [61], Rizwan et al. (2015) [62], Belletante et al. (2016) [63], Hernández-Calderon et al. (2016) [64], and Martin et al. (2016) [65]. This methodology allows assessing different optimization objectives such as maximizing yield and profit and minimizing costs, wastes, fixed costs, and environmental impact.

The main steps of the optimization methodology of superstructures are mentioned as follows [46], [47]. The majority of the authors follow this order with small changes.

1. *To define the scope and complexity of the problem.* Process systems engineering (PSE) approach is needed to solve the complexity of the product allocation problem in the processing facilities. The complexity of the problem does not allow solving through heuristics or thumb rules. The PSE approach uses techniques such as process integration and mathematical optimization to ensure a target approach combining simulation and experimental works. Besides, the PSE can ensure a high economic and social benefit reducing the use of feedstocks and energy. The optimization methodology involves the

superstructure concept, where all possible transformation routes to obtain a product set are considered, maximizing the net present value and minimizing the environmental impact [46,47].

A mathematical optimization based on the framework is carried out to fulfill the objectives mentioned previously. This allows the inclusion of techno-economic aspects, information from experimental studies, modeling, and simulation. The framework answers to issues such as product prices, product amount, best configuration, and optimal conditions, among others.

2. *To integrate modeling and experimental data.* A strict library composed by simulation models of transformation routes and a database of yields is built in this step. Experimental results are the base to validate the simulations of processes using a commercial software package [46,47].

   The proposal is to assess stand-alone processes (i.e., simulation models) that compose the global biorefinery considering these substeps: (i) extracting data and knowledge from literature and experiments; (ii) simulating models using Aspen Plus, HYSYS, and/or Pro II, among others; (iii) validating models performance; (iv) integrating processes using tools such as pinch analysis, thermal management, and resource conservation strategies in order to optimize the simulation models; (v) optimizing simulation models considering a minimum use of utilities, a maximum use of resources, and a reduction in the environmental impact; and (vi) obtaining economic data through cost estimation software and references, suppliers data, etc., and environmental data through U.S. EPA WAR algorithm, SimaPro, GREENSCOPE, databases, etc. In the validation step (i.e., substep iii), when a process requires a solvent, molecular design techniques should be employed in order to select the best option from an environmental and safety point of view [46,47].

   The abovementioned six substeps are the base to generate a compact model library and performance database (i.e., economic and environmental potential) and a superstructure of transformation routes (i.e., a tree with all optimized models). This approach promotes a constant upgrade or a new inclusion of models, without changing the methodology.

3. *To optimize the biorefinery.* The optimization framework considers a combination between the models library and the economic performance with a numerical solver, namely, a mixture of the following factors: (i) objectives of process design, (ii) processing superstructures, (iii) constraints, (iv) performance of the framework database (economic potential), and (v) numerical solver routines such as Mixed Integer Linear Programming (MILP) or Mixed Integer Nonlinear Programming (MINLP) [46,47].

The goal of the optimization framework is to identify the solutions capable of maximizing the economic performance, and when the solution is positive, it passes to the assessment of the environmental performance. If the solution fulfills the environmental constraints, an optimal process scheme has been identified. If the environmental solution is negative, constraints are relaxed until achieving acceptable environmental results.

1.3.3 Early-Stage Method

Initially, the early-stage sustainability assessment method uses a multiobjective decision framework for the preliminary assessment of chemical processes at the laboratory stage. This method allows a revision of chemical processes in a wide sustainability context because a multi-criteria approach is used with a combination of economic, environmental, safety, and health indicators [66]. However, this approach is modified to bio-based processes and considers the characteristics such as biomass pretreatment, co-product number, risks, and comparison with the equivalent petrochemical part [67–69].

This approach applied to the design of biorefineries appears as an alternative to known methodologies such as (i) optimization of superstructures and (ii) process engineering and conceptual design of processes [68,70]. The goal of the early-stage sustainability assessment method focused on biorefineries is to assess systematically the biorefinery products to identify those with the best potential from a sustainability point of view at early design stages [68,70].

The method is used to determine the sustainability of catalytic and chemical processes and to identify their strengths and weaknesses. Three sustainability indicators are considered: (i) economic constraint, (ii) energy-related impact of raw materials, and (iii) process complexity. These indicators are grouped in a single index with the help of weighting factors to have a fast comparison and selection of the process [68,70]. The method applied to the design of biorefineries considers five steps that are mentioned as follows [70]:

1. *Modification of the method.* Initially, the method is applied to chemical processes and stand-alone bio-based processes considering five indicators. Then, it is applied to the design of biorefineries (i.e., a combination of chemical, thermochemical, mechanical, and biochemical processes) considering three indicators.

2. *Definition of the study cases.* The study cases are defined according to the literature revision and the selection of chemicals, derivatives, and platforms.

3. *Data collection.* The data are collected based on the literature, databases, and analyses of stand-alone processes.

4. *Application of the method.* First, the sustainability scores are calculated based on the stand-alone bioprocesses, and then a comparison with the reference system is made. The method includes analysis of sensitivity and scenarios.

5. *Analysis of results.* The analysis of results considers the classification of derivatives, the comparison of platforms, and the identification of promising derivatives.

1.4 Conclusions

As described in this chapter, the relationship between the concept (the definition), the raw materials (mainly lignocellulosics as a world tendency), the technologies (to improve the integral and efficient use of the raw materials), and the design methods demonstrated that biorefinery is a complex task. Then the design, analysis, and implementation should be done using the most powerful tools and strategies which are described in this book.

References

1. IEA Bioenergy Task42, "Biorefineries: Adding value to the sustainable utilisation of biomass," IEA Bioenergy, vol. 1, pp. 1–16, 2009.
2. B. Kamm and M. Kamm, "International biorefinery systems," *Pure Appl. Chem.,* vol. 79, pp. 1983–1997, 2007.
3. B. Kamm, M. Kamm, and J. Venus, "Chapter 6: Principles of biorefineries - The role of biotechnology the example lactic acid fermentation," in *Trends in Biotechnology Research*, pp. 199–223 Hauppauge: Nova Science Publishers, Inc., 2006.
4. R. Van Ree and B. Annevelink, "Status report biorefinery 847," 2007.
5. H.-J. Huang, S. Ramaswamy, U. W. Tschirner, and B. V. Ramarao, "A review of separation technologies in current and future biorefineries," *Sep. Purif. Technol.,* vol. 62, no. 1, pp. 1–21, Aug. 2008.
6. F. Cherubini and S. Ulgiati, "Crop residues as raw materials for biorefinery systems—A LCA case study," *Appl. Energy,* vol. 87, pp. 47–57, 2010.
7. L. Axelsson, M. Franzén, M. Ostwald, G. Berndes, G. Lakshmi, and N. H. Ravindranath, "Perspective: Jatropha cultivation in southern India: Assessing farmers' experiences," *Biofuels, Bioprod. Biorefining,* vol. 6, pp. 246–256, 2012.
8. J. Moncada, V. Aristizábal, and C. A. Cardona, "Design strategies for sustainable biorefineries," *Biochem. Eng. J.,* vol. 116, pp. 122–134, 2016.
9. C. A. Cardona, V. Aristizábal, and J. C. Solarte Toro, "Chapter 2: Improvement of palm oil production for food industry through biorefinery concept," in *Advances in Chemistry Research*, J. C. Taylor Ed., Vol. 32 New York: Nova Science Publishers, 2016, pp. 37–72.

10. N. Q. Diep, K. Sakanishi, N. Nakagoshi, S. Fujimoto, T. Minowa, and X. D. Tran, "Biorefinery: Concepts, current status, and development trends," *Int. J. Biomass Renew.*, vol. 1, no. January, pp. 1–8, 2012.

11. A. Sonnenberg, J. Baars, and P. Hendrickx, "Brochure IEA Bioenergy—Task 42 Biorefinery," 2013.

12. L. R. Lynd, C. Wyman, M. Laser, D. Johnson, and R. Landucci, Strategic Biorefinery Analysis: Analysis of Biorefineries. Golden, CO: National renewable energy laboratory—NREL, 2005.

13. F. Cherubini, "The biorefinery concept: Using biomass instead of oil for producing energy and chemicals," *Energy Convers. Manag.*, vol. 51, no. 7, pp. 1412–1421, 2010.

14. A. R. Morais and R. Bogel-Lukasik, "Green chemistry and the biorefinery concept," *Sustainable Chemical Processes*, vol. 1, pp. 1–18, 2013.

15. Bioenergy IEA, "Task 42—Bio-based chemicals. Value added products from biorefineries," *IEA Bioenergy*, pp. 1–36, 2012.

16. J. Moncada, M. M. El-Halwagi, and C. A. Cardona, "Techno-economic analysis for a sugarcane biorefinery: Colombian case," *Bioresour. Technol.*, vol. 135, pp. 533–543, 2013.

17. J. A. Posada, L. E. Rincón, and C. A. Cardona, "Design and analysis of biorefineries based on raw glycerol: Addressing the glycerol problem," *Bioresour. Technol.*, vol. 111, pp. 282–293, May 2012.

18. A. Demirbas, "Biorefineries: Current activities and future developments," *Energy Convers. Manag.*, vol. 50, pp. 2782–2801, 2009.

19. F. Carvalheiro, L. C. Duarte, and F. M. Gírio, "Hemicellulose biorefineries: A review on biomass pretreatments," *J. Sci. Ind. Res. (India)*, vol. 67, pp. 849–864, 2008.

20. Á. Németh and Z. Kaleta, "Complex utilization of dairy waste (whey) in biorefinery," *Wseas Trans. Environ. Dev.*, vol. 11, pp. 80–88, 2015.

21. V. Aristizábal, Á. Gómez, and C. A. Cardona, "Biorefineries based on coffee cut-stems and sugarcane bagasse: Furan-based compounds and alkanes as interesting products," *Bioresour. Technol.*, vol. 196, pp. 480–489, 2015.

22. J. Moncada, J. Tamayo, and C. A. Cardona, "Evolution from biofuels to integrated biorefineries: Techno-economic and environmental assessment of oil palm in Colombia," *J. Clean. Prod.*, vol. 81, pp. 51–59, 2014.

23. A. A. Koutinas, R. Wang, G. M. Campbell, and C. Webb, "A whole crop biorefinery system: A closed system for the manufacture of non-food products from cereals," in *Biorefineries-Industrial Processes and Products: Status Quo and Future Directions*, vol. 1, pp. 165–191. Weinheim: Wiley-VCH, 2008.

24. V. Aristizábal, C. A. García, and C. A. Cardona, "Integrated production of different types of bioenergy from oil palm through biorefinery concept," *Waste Biomass Valori.*, vol. 7, no. 4, pp. 737–745, 2016.

25. C. A. M. Silva, R. M. Prunescu, K. V. Gernaey, G. Sin, and R. A. Diaz-Chavez, "Biorefinery Sustainability Analysis," in *Biorefineries*, M. Rabaçal, A. F. Ferreira, C. A. M. Silva, and M. Costa, Eds., pp. 161–200. Lecture Notes in Energy, vol 57. Cham: Springer, 2017.

26. J. Moncada, I. Vural Gursel, W. J. J. Huijgen, J. W. Dijkstra, and A. Ramírez, "Techno-economic and ex-ante environmental assessment of C6 sugars production from spruce and corn. Comparison of organosolv and wet milling technologies," *J. Clean. Prod.*, vol. 170, pp. 610–624, 2018.

27. Office of Energy Efficiency & Renewable Energy. U.S. Department of Energy, "Integrated Biorefineries," 2017. [Online]. Available: www.energy.gov/eere/bio-energy/integrated-biorefineries (Accessed 03 Feb. 2017.)

28. R. van Ree and A. van Zeeland, "IEA Bioenergy—Task42 Biorefining," 2014.

29. M. Galbe and G. Zacchi, "Pretreatment of lignocellulosic materials for efficient bioethanol production," *Adv. Biochem. Eng. Biotechnol.*, vol. 108, pp. 41–65, 2007.

30. P. Kumar, D. M. Barrett, M. J. Delwiche, and P. Stroeve, "Methods for pretreatment of lignocellulosic biomass for efficient hydrolysis and biofuel production," *Ind. Eng. Chem.*, vol. 48, pp. 3713–3729, 2009.

31. P. Harmsen, W. Huijgen, L. López, and R. Bakker, "Literature review of physical and chemical pretreatment processes for lignocellulosic biomass," in *Food and Biobased Research*, pp. 1–49, 2010.

32. A. T. W. M. Hendriks and G. Zeeman, "Pretreatments to enhance the digestibility of lignocellulosic biomass," *Bioresour. Technol.*, vol. 100, no. 1, pp. 10–18, 2009.

33. V. B. Agbor, N. Cicek, R. Sparling, A. Berlin, and D. B. Levin, "Biomass pretreatment: Fundamentals toward application," *Biotechnol. Adv.*, vol. 29, pp. 675–685, 2011.

34. C. A. Cardona, L. F. Gutierrez, and O. J. Sánchez, "Chapter 4: Feedstock conditioning and pretreatment," in *Process Synthesis for Fuel Ethanol Production*, pp. 77–113. Nova Publishers, Inc., 2010.

35. M. E. Zakrzewska, E. Bogel-łukasik, and R. Bogel-łukasik, "Ionic liquid-mediated formation of 5-Hydroxymethylfurfural a promising biomass-derived building block," *Am. Chem. Soc.*, vol. 111, pp. 397–417, 2011.

36. A. Martinez, M. E. Rodriguez, M. L. Wells, S. W. York, J. F. Preston, and L. O. Ingram, "Detoxification of dilute acid hydrolysates of lignocellulose with lime," *Biotechnol. Prog.*, vol. 17, no. 2, pp. 287–293, 2001.

37. J. Clark and F. Deswarte, "Chapter 1: The biorefinery concept—An integrated approach," in *Introduction to Chemicals from Biomass*, 2nd ed., 2008, pp. 1–29.

38. F. Cherubini et al., "Toward a common classification approach for biorefinery systems," *Biofuels, Bioprod. Biorefining*, vol. 5, pp. 1–13, 2009.

39. V. Aristizábal Marulanda, C. D. Botero G, and C. A. Cardona A., "Chapter 4: Thermochemical, biological, biochemical, and hybrid conversion methods of bio-derived molecules into renewable fuels," in *Advanced Bioprocessing for Alternative Fuels, Biobased Chemicals, and Bioproducts*, 2017, p. In press. Elsevier.

40. R. Smith, *Chemical Process Design and Integration*. New York, NY: McGraw-Hill. 2005.

41. J. M. Douglas, *Conceptual Design of Chemical Processes*. New York, NY: McGraw-Hill. 1988.

42. H. Cabezas, J. C. Bare, and S. K. Mallick, "Pollution prevention with chemical process simulators: The generalized waste reduction (WAR) algorithm," *Comput. Chem. Eng.*, vol. 21, pp. S305–S310, 1997.

43. D. M. Young and H. Cabezas, "Designing sustainable processes with simulation: the waste reduction (WAR) algorithm," *Comput. Chem. Eng.*, vol. 23, pp. 1477–1491, 1999.

44. D. Young, R. Scharp, and H. Cabezas, "The waste reduction (WAR) algorithm: environmental impacts, energy consumption, and engineering economics," *Waste Manag.*, vol. 20, pp. 605–615, 2000.

45. J. Moncada Botero, "Design and evaluation of sustainable biorefineries from feedstock in tropical regions," Universidad Nacional de Colombia. Departamento de Ingeniería Química. Master Thesis, 2012.
46. N. Sammons Jr, M. Eden, W. Yuan, H. Cullian, and B. Aksoy, "A flexible framwork for optimal biorefinery product allocation," *Am. Inst. Chem. Eng.*, vol. 4, no. 2, pp. 349–354, 2007.
47. N. E. Sammons, W. Yuan, M. R. Eden, B. Aksoy, and H. T. Cullinan, "Optimal biorefinery product allocation by combining process and economic modeling," *Chem. Eng. Res. Des.*, vol. 86, no. 7, pp. 800–808, 2008.
48. E. Zondervan, M. Nawaz, A. B. De Haan, J. M. Woodley, and R. Gani, "Optimal design of a multi-product biorefinery system," *Comput. Chem. Eng.*, vol. 35, pp. 1752–1766, 2011.
49. M. O. Bertran, R. Frauzem, A. S. Sanchez-Arcilla, L. Zhang, J. M. Woodley, and R. Gani, "A generic methodology for processing route synthesis and design based on superstructure optimization," *Comput. Chem. Eng.*, vol. 106, pp. 892–910, 2017.
50. J. E. Santibañez-Aguilar, J. B. González-Campos, J. M. Ponce-Ortega, M. Serna-González, and M. M. El-Halwagi, "Optimal planning of a biomass conversion system considering economic and environmental aspects," *Ind. Eng. Chem. Res.*, vol. 50, pp. 8558–8570, 2011.
51. S. Giarola, A. Zamboni, and F. Bezzo, "Spatially explicit multi-objective optimisation for design and planning of hybrid first and second generation biorefineries," *Comput. Chem. Eng.*, vol. 35, pp. 1782–1797, 2011.
52. J. M. Ponce-Ortega, V. Pham, M. M. El-Halwagi, and A. A. El-Baz, "A disjunctive programming formulation for the optimal design of biorefinery configurations," *Ind. Eng. Chem. Res.*, vol. 51, pp. 3381–3400, 2012.
53. V. Pham and M. El-Halwagi, "Process synthesis and optimization of biorefinery configurations," *AIChE J.*, vol. 58, pp. 1212–1221, 2012.
54. F. You, L. Tao, D. J. Graziano, and S. W. Snyder, "Optimal design of sustainable cellulosic biofuels sypply chains: Multiobjective optimization coupled with life cycle assessment and input-output analysis," *AIChE J.*, vol. 58, pp. 1157–1180, 2012.
55. M. El-Halwagi, C. Rosas, J. M. Ponce-Ortega, A. Jiménez-Gutierrez, M. Sam Mannan, and M. M. El-Halwagi, "Multi-objective optimization of biorefineries with economic and safety objectives," *AIChE J.*, vol. 59, pp. 2427–2434, 2013.
56. B. Wang, B. H. Gebreslassie, and F. You, "Sustainable design and synthesis of hydrocarbon biorefinery via gasification pathway: Integrated life cycle assessment and technoeconomic analysis with multiobjective superstructure optimization," *Comput. Chem. Eng.*, vol. 52, pp. 55–76, 2013.
57. K. Tong, J. Gong, D. Yue, and F. You, "Stochastic programming approach to optimal design and operations of integrated hydrocarbon biofuel and petroleum supply chains," *ACS Sustain. Chem. Eng.*, vol. 2, pp. 49–61, 2014.
58. A. Kelloway and P. Daoutidis, "Process synthesis of biorefineries: Optimization of biomass conversion to fuels and chemicals," *Ind. Eng. Chem. Res.*, vol. 53, pp. 5261–5273, 2014.
59. J. Gong and F. You, "Optimal design and synthesis of algal biorefinery processes for biological carbon sequestration and utilization with zero direct greenhouse gas emissions: MINLP model and global optimization algorithm," *Ind. Eng. Chem. Res.*, vol. 53, pp. 1563–1579, 2014.

60. J. E. Santibañez-Aguilar, J. B. González-Campos, J. M. Ponce-Ortega, M. Serna-González, and M. M. El-Halwagi, "Optimal planning and site selection for distributed multiproduct biorefineries involving economic, environmental and social objectives," *J. Clean. Prod.*, vol. 65, pp. 270–294, 2014.

61. A. Geraili and J. A. Romagnoli, "A multiobjective optimization framewrok for design of integrated biorefineries under uncertainty," *AIChE J.*, vol. 61, pp. 3208–3222, 2015.

62. M. Rizwan, J. H. Lee, and R. Gani, "Optimal design of microalgae-based biorefinery: Economics, opportunities and challenges," *Appl. Energy*, vol. 150, pp. 69–79, 2015.

63. S. Belletante, L. Montastruc, S. Negny, and S. Domenech, "Optimal design of an efficient, profitable and sustainable biorefinery producing acetone, butanol and ethanol: Influence of the in-situ separation on the purification structure," *Biochem. Eng. J.*, vol. 116, pp. 195–209, 2016.

64. O. M. Hernández-Calderón et al., "Optimal design of distributed algae-based biorefineries using CO_2 emissions from multiple industrial plants," *Ind. Eng. Chem. Res.*, vol. 55, pp. 2345–2358, 2016.

65. M. Martín and I. E. Grossmann, "Optimal production of furfural and DMF from algae and switchgrass," *Ind. Eng. Chem. Res.*, vol. 55, pp. 3192–3202, 2016.

66. H. Sugiyama, U. Fischer, K. Hungerbühler, and M. Hirao, "Decision framework for chemical process design including different stages of environmental, health and safety assessment," *AIChE J.*, vol. 54, pp. 1037–1053, 2008.

67. A. D. Patel, K. Meesters, H. den Uil, E. de Jong, K. Blok, and M. K. Patel, "Sustainability assessment of novel chemical processes at early stage: Application to biobased processes," *Energy Environ. Sci.*, vol. 5, no. 9, p. 8430, 2012.

68. J. A. Posada, A. D. Patel, A. Roes, K. Blok, A. P. C. Faaij, and M. K. Patel, "Potential of bioethanol as a chemical building block for biorefineries: Preliminary sustainability assessment of 12 bioethanol-based products," *Bioresour. Technol.*, vol. 135, pp. 490–499, 2013.

69. A. D. Patel, K. Meesters, H. den Uil, E. de Jong, E. Worrell, and M. K. Patel, "Early-stage comparative sustainability assessment of new bio-based processes," *ChemSusChem*, vol. 6, no. 9, pp. 1724–1736, Oct. 2013.

70. J. Moncada, J. A. Posada, and A. Ramirez, "Early sustainability assessment for potential configurations of integrated biorefineries. Screening platform chemicals," *Biofuels, Bioprod. Biorefining*, vol. 9, pp. 722–748, 2015.

2

Biorefinery Feedstocks, Platforms, and Products

Biorefineries are based on four important pillars or components: feedstocks, platforms, products, and conversion routes, which are connected together in a sequential and hierarchical way. In Chapter 1, the conversion routes were discussed as one of the main topics, where the majority of technologies for pretreatment, reaction, and purification were considered and described. In this chapter, an overview of all bio-based feedstocks and their respective classification is presented. Then, the platforms as articulating shaft in a biorefinery are indicated due to the feedstocks are transformed first into platforms and then into products through different transformation routes. Finally, the main products of the biorefineries are discussed and some examples are presented.

2.1 Biomass Feedstocks

In a conventional way, feedstock can be defined as the raw material to feed or supply to a process or machine in order to be transformed or converted into a good or service. In the context of bio-based processes and biorefineries, the main feedstock is the biomass, which refers to all existing vegetable material (plants and algae), but in many cases, it can also be considered any material from animal origin such as whey and fats [1]. The Office of Energy Efficiency and Renewable Energy of the U.S. Energy Department defines a feedstock as "a biological and renewable material that can be used directly as fuel or transformed to other fuel form or energy product" [2]. However, this definition is only related to energy, which today is not totally true. In addition, the biomass feedstocks can be used to obtain food, biopolymers [3,4], biochemicals [5], and biofertilizers [6], among others. The National Renewable Energy Laboratory defines biomass in detail, making a brief classification according to its origin: "The biomass includes herbaceous and woody energy crops, agricultural food and feed crops, agricultural crop wastes and residues, wood wastes and residues, aquatic plants, and other waste materials including some municipal wastes. Biomass is a very heterogeneous and chemically complex renewable resource." In decreasing order of abundance, the elements in biomass are commonly C, O, H, N, Ca, K, Si, Mg, Al, S, Fe, P, Cl, Na, Mn, and Ti [7]. The main structural ingredients of biomass are cellulose, hemicellulose, lignin, starch, extractives, and moisture. The biomass composition depends on the factors such as cultivation process, growing conditions (geographic location, climate, seasons, water, pH, and nutrients), dosing of fertilizers and pesticides, harvesting time and collection techniques, transport, and storage conditions [7].

Another important definition is given by the U.S. Energy Information Administration (EIA), which considers the biomass as "an organic non-fossil material of biological origin constituting a renewable energy source," and in this sense, EIA presents a more specific term: biomass waste, whose definition is "Organic non-fossil material of biological origin that is a byproduct or a discarded product. Biomass waste includes municipal solid waste from biogenic sources, landfill gas, sludge waste, agricultural crop byproducts, straw, and other biomass solids, liquids, and gases" [8]. In the past few years, there has been a growing interest in biomass used as an alternative to nonrenewable feedstocks (e.g., petroleum) to produce biochemicals and biofuels through green processes. Biomass has interesting features such as renewable energy supply, high availability, relatively low cost, high influence on the revitalization of rural areas and economy, and environmental considerations that involve a clean feedstock due to the negligible content of sulfur, nitrogen, and ash; CO_2 neutral conversion; and climate change benefits [7,9,10]. The diversity in biomass composition is an additional characteristic that allows its conversion into an important range of products.

Figure 2.1 shows the classification of biomass feedstocks in a biorefinery taking into account their origin: dedicated energy crops and biomass residues [11,12]. These groups are divided into subgroups such as sugar crops, starch crops, and lignocellulosic crops. Dedicated energy crops are plants that do not produce food and are cultivated mainly for energy production, growing on marginal soils and nonarable lands, and do not need large amounts of water and present high yields per hectare. Biomass residues are by-products from the harvesting or processing of biomass that presents a considerable energy potential. Both groups can be obtained from four productive sectors: agriculture, forestry, industry, and aquaculture [11,13]. It is important to note again that the concept of "energy potential" should be considered just as a result of the past 40 years boom on bioenergy as a main product from biomass. Hereafter in this book, the "energy potential," which is one of the most important characteristics of biomass, is modified to "biorefinery potential" in all the previous definitions of biomass. The biorefinery potential establishes the potential to produce not only energy but also chemicals, and food and feed ingredients from biomass in an integrated way.

From the biorefinery point of view, the biomass feedstocks can also be classified into four main generations: the first generation (1G) considering food crops [13,14]; the second generation (2G) based on the residues of food crops, dedicated energy crops, and nonedible crops [13,15,16]; the third generation

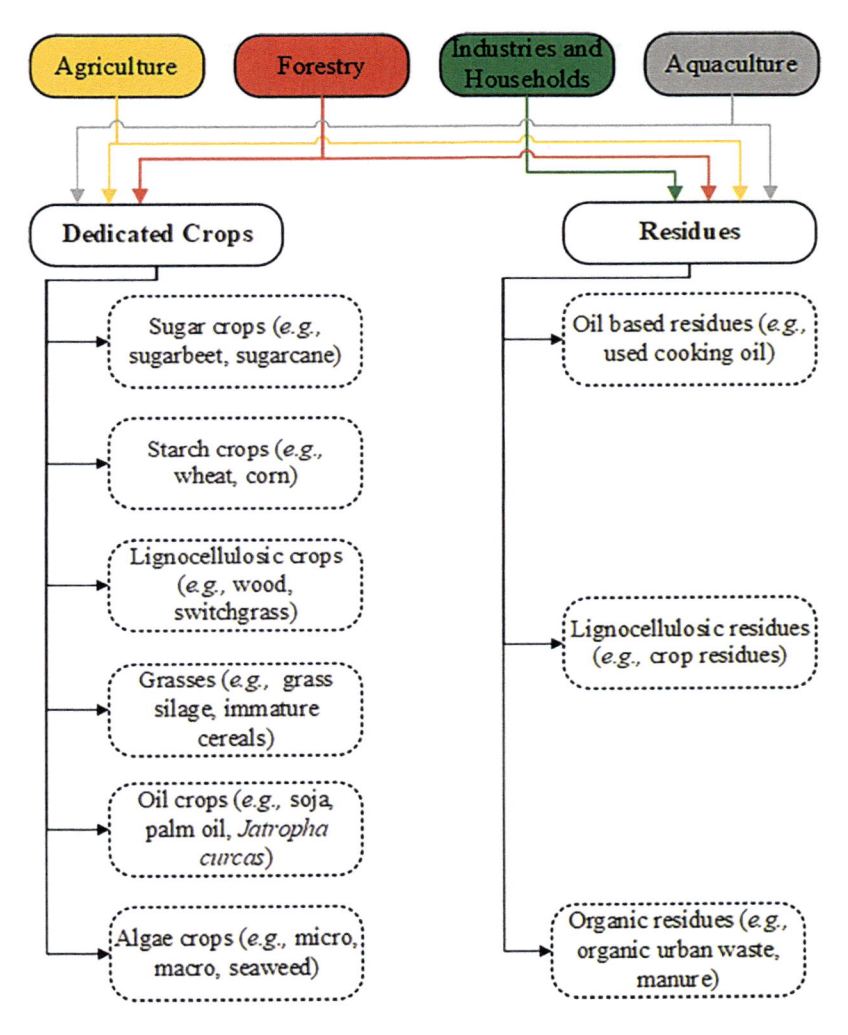

FIGURE 2.1
Classification of the feedstocks according to their origin. (Adapted from [12].)

(3G) considering the use of algae as feedstocks [17–19]; and the fourth generation (4G) using captured CO_2 mainly [17,20]. This classification comes mainly from the same strategy used for biofuels, being the first products obtained from biomass as stand-alone processes [13,21,22]. The consolidation of biofuels as a well-known and established industry contributes to the categorization of feedstocks according to their characteristics, and these same concepts can be applied in biorefineries. In Sections 2.1.1–2.1.4, a description of each generation of feedstocks and some examples of studied bio-based processes and biorefineries are presented.

2.1.1 1G Feedstocks

Edible and agricultural crops such as vegetable oils, animal fats, cereals, sugarcane, sugar beet, and starches are considered, among others, as 1G feedstocks. These primary sources are produced by photosynthesis and are harvested/collected directly from the land. This type of biomass presents social, economic, and environmental challenges because these are derived from food crop feedstocks [20,21,23]. Its use can lead to increase food prices, and also to create pressure on land use for direct competition with food, arable land, and freshwater. These factors along with deficiencies in policies for the agriculture sector can make it unlikely to be totally sustainable. The limitations in the generation of feedstocks have been partially overcome through the alternative materials so-called 2G, 3G, and 4G feedstocks [13]. However, the consideration of this type of feedstocks in bio-based processes presents some advantages such as greenhouse gas (GHG) savings, and conventional and low-cost conversion technologies due to the presence of easily digestible sugars [13,17]. Table 2.1 shows some examples of bioprocesses and biorefineries considering 1G feedstocks, with their platforms, technologies, and products, remarking that one raw material can provide different configurations or scenarios.

2.1.2 2G Feedstocks

2G feedstocks are the by-products of harvesting or processing 1G feedstocks. This generation involves nonedible crops and residues (e.g., jatropha, wood waste, waste cooking oil, forestry residues) [13,16]. The 2G feedstocks have advantages over 1G feedstocks such as high availability, GHG savings, use of nonarable lands, reduction of water demand, and low competition with food crops [17,22]. However, each biorefinery should be studied in detail to avoid new competition of the 2G feedstocks with fertilizers (e.g., leaves that remain in the field after harvesting can act as the source of nutrients for the soil) or feed security (e.g., most of the produced feed for animals is based on residual biomass). It is important to note that these two examples can finally affect the food security making 2G raw materials not so competitive against 1G feedstocks. The 2G sources can overcome social, economic, and environmental challenges without hampering the food cost and creating pressure on land use. Some limitations of these feedstocks are linked to the need to develop advanced technologies to pretreat these materials in order to access to simple sugars and convert them to biofuels and biochemicals in a cost-effective way [17]. Table 2.2 shows the study cases of bioprocesses and biorefineries using 2G feedstocks, with their platforms, technologies, and products.

The majority of these feedstocks are lignocellulosic materials, which are classified into three groups: (i) agricultural residues, (ii) forest residues, and (iii) energy crops [16]. Agricultural residues (i.e., husk, bagasse, and stems, among others) come from crops such as corn, rice, coffee, sugarcane,

TABLE 2.1

Examples of Bioprocesses and Biorefineries Considering 1G Feedstocks (Based on the Literature)

Feedstock	Type of Process	Platforms	Technologies	Products	Remarks	Reference
Theobroma grandiflorum (Copoazu fruit: pulp, seeds, and peel)	1G biorefinery	Plant-based oil, C5/C6 sugars and lignin	*Pretreatment stage.* Drying, milling, cracking. *Reaction stage.* Pasteurization, fermentation using *Zymomonas mobilis* and *Cupriavidus necator*, anaerobic digestion, and gasification. *Separation stage.* Supercritical fluid extraction, nanofiltration membrane, steam distillation, solvent extraction, distillation, molecular sieves, centrifugation, and gas turbine	Pasteurized pulp, antioxidant extract, biofertilizer, biogas, oil seed, essential oil, ethanol, and polyhydroxybutyrate (PHB)	Three scenarios were assessed from a techno-economic and an environmental point of view taking into account different integration levels—scenario 1: nonintegrated; scenario 2: fully energy integrated; and scenario 3: fully integrated and cogeneration system. The total economic margin in the best configuration (scenario 3) was 13.21%. The potential environmental impact decreased not only due to the energy integration but also due to the use of solid residues for energy cogeneration to produce steam.	[24]
Theobroma bicolor (Makambo fruit: pulp, seeds, and peel)	1G biorefinery	Organic solutions and lignin	*Pretreatment stage.* Drying, milling, pressing. *Reaction stage.* Anaerobic digestion. *Separation stage.* Supercritical fluid extraction, evaporation, ultrafiltration	Pasteurized pulp, butter and residual cake (paste as a substitute for cacao), extract of phenolic compounds, biogas, and biofertilizer	Two biorefinery systems were analyzed in order to establish their potential as an economic alternative. Scenario 1 considered the obtaining of pasteurized pulp, butter, and residual cake, and scenario 2 taken into account these	[25]

(Continued)

TABLE 2.1 (*Continued*)

Examples of Bioprocesses and Biorefineries Considering 1G Feedstocks (Based on the Literature)

Feedstock	Type of Process	Platforms	Technologies	Products	Remarks	Reference
					products plus extract of phenolic compounds, biogas, and biofertilizer. The results showed that the second scenario has the best performance from the economic and environmental perspectives.	
Residual banana (pulp and peel)	Stand-alone bioprocess and 1G biorefinery	Organic solutions and C6 sugar	*Pretreatment stage.* Liquefaction, acid hydrolysis, *Reaction stage.* Enzymatic hydrolysis and fermentation using *Burkholderia sacchari* and *Pichia stipitis* *Separation stage.* Evaporation, crystallization, distillation, molecular sieves, centrifugation, evaporation, and spray drying	Ethanol, glycose, and polyhydroxy-butyric acid (PHB)	Three scenarios were analyzed— scenario 1: PHB stand-alone bioprocess; scenario 2: PHB, glucose, and ethanol biorefinery; and scenario 3: biorefinery plus mass and energy integration. The results showed that the energetic integration can reduce up to 30.6% the global energy requirements of the process, and the mass integration allowed a 35% water saving.	[26]

(*Continued*)

TABLE 2.1 (Continued)

Examples of Bioprocesses and Biorefineries Considering 1G Feedstocks (Based on the Literature)

Feedstock	Type of Process	Platforms	Technologies	Products	Remarks	Reference
Oil palm	Two 1G biorefineries	Plant-based oil, C6/C5 sugars, syngas, and lignin	*Pretreatment stage.* Milling, acid hydrolysis, *Reaction stage.* Enzymatic hydrolysis, fermentation using *Z. mobilis* and *Escherichia coli*, transesterification reaction, neutralization, gasification. *Separation stage.* Distillation, molecular sieves, centrifugation, liquid–liquid extraction, gas turbine	Biorefinery 1: biodiesel and bioethanol Biorefinery 2: biodiesel, palm olein, methanol, heat, and power	Two biorefineries were designed. The potential income (total sales/total production cost ratio) for the biorefineries 1 and 2 were 1.88 and 3.33, respectively, and the potential environmental impacts per ton were 90 and 240, respectively.	[27]
Sugarcane	Stand-alone process	Organic solution (i.e., cane juice)	*Pretreatment stage.* Milling *Reaction stage.* Clarification, neutralization, fermentation using *Z. mobilis* and *S. cerevisiae*.	Ethanol	The ethanol production was carried out with several technological configurations, considering the economic and environmental criteria. The configuration that considered the continuous cultivation of *Z. mobilis* with cane juice as substrate, including cell	[28]

(Continued)

TABLE 2.1 (*Continued*)

44

Examples of Bioprocesses and Biorefineries Considering 1G Feedstocks (Based on the Literature)

Feedstock	Type of Process	Platforms	Technologies	Products	Remarks	Reference
			Separation stage. Distillation, molecular sieves		recycling and ethanol dehydration by molecular sieves, was the most suitable.	
Sugarcane	1G biorefinery	Organic solution (i.e., cane juice and molasses), C6/C5 sugars and lignin	*Pretreatment stage.* Milling, acid hydrolysis. *Reaction stage.* Clarification, fermentation using *Z. mobilis, S. cerevisiae, C. necator,* and *Felicia anassa,* detoxification reaction, gasification *Separation stage.* Evaporation, spray drying, crystallization, distillation, molecular sieves, solvent extraction, ultrafiltration, gas turbine.	Sugar, ethanol, PHB, anthocyanins, and electricity	Two scenarios were considered for the production of sugar, fuel ethanol, PHB, anthocyanins, and electricity. Scenario 1 used for ethanol production, 70% of glucose from hydrolysis of molasses and 47.5% of cane juice and for PHB production, 30% of glucose from hydrolysis of molasses, and 47.5% of cane juice. Scenario 2 used for ethanol production, 70% of glucose from cellulose hydrolysis from cane bagasse and molasses, 100% of xylose from hemicellulose hydrolysis from bagasse and for PHB production, and 30% of glucose from cellulose hydrolysis and molasses. These scenarios considered different conversion pathways as a function of feedstock distribution and technologies. The base case that produces sugar, ethanol, and electricity was compared with these scenarios.	[29]

(Continued)

TABLE 2.1 (*Continued*)

Examples of Bioprocesses and Biorefineries Considering 1G Feedstocks (Based on the Literature)

Feedstock	Type of Process	Platforms	Technologies	Products	Remarks	Reference
					The results showed that the configuration with the best economic, environmental, and social performance is the one that considers fuel ethanol and PHB production from the combined cane bagasse and molasses.	
Oil palm	Stand-alone bioprocess and 1G biorefinery	Plant-based oil and C6/C5 sugars	*Pretreatment stage.* Acid hydrolysis *Reaction stage.* Pre-esterification, transesterification and neutralization, enzymatic hydrolysis, detoxification, fermentation using Z. *mobilis* and *C. necator* *Separation stage.* Washing, vacuum flash, distillation, molecular sieves, evaporation, spray drying	Biodiesel, ethanol, and PHB	Four scenarios were assessed—scenario 1: stand-alone biodiesel process, scenario 2: biodiesel process with mass integration; scenario 3: biorefinery to produce biodiesel, ethanol, and PHB; and scenario 4: biorefinery plus mass integration. Scenario 1 presented the best economic and environmental performance with an economic margin of 64.5% and a carbon footprint of 0.51 t CO_2eq/m^3 of biodiesel	[30]

TABLE 2.2

Examples of Bioprocesses and Biorefineries Considering 2G Feedstocks (Based on the Literature)

Feedstock	Type of Process	Platforms	Technologies	Products	Remarks	Reference
Orange peel	2G biorefinery	Plant-based oil and lignin	*Pretreatment stage.* Milling, acid hydrolysis *Reaction stage.* Gasification, dehydrogenation reaction *Separation stage.* Solvent extraction, drying, decantation, distillation, evaporation, gas turbine	*p*-Cymene, hydrogen, and pectin	Two scenarios were evaluated (with and without electricity generation) obtaining *p*-cymene (97% purity) and pectin (81% purity) with productivities of 9.22 and 42.63 kg/h, respectively. From the techno-economic analysis, the best scenario did not consider the electricity generation with production costs of 5.27 and 3.53 USD/kg of *p*-cymene and pectin, respectively.	[31]
Spent pulp of Colombian Andes Berry (*Rubus glaucus* Benth.)	2G biorefinery	C5/C6 sugars and lignin	*Pretreatment stage.* Milling, drying, acid hydrolysis. *Reaction stage.* Detoxification, enzymatic hydrolysis, fermentation using *Candida guilliermondii* and *S. cerevisiae* and gasification *Separation stage.* Supercritical extraction, spray drying, evaporation, crystallization, distillation, gas turbine	Phenolic compounds, ethanol, xylitol and electricity	Four scenarios including mass and heat integrations as well as cogeneration system were considered. The economic analysis revealed that the level of integration in the biorefinery significantly affected the total production cost. The cost of supplies (enzymes and reagents) had the most significant impact, representing between 46.72% and 58.95% of the total cost of the biorefinery.	[32]
Plantain pseudostem and rice husk	2G biorefinery	C5/C6 sugars and lignin	*Pretreatment stage.* Drying, milling, acid hydrolysis *Reaction stage.* Detoxification, enzymatic hydrolysis, fermentation using *S. cerevisiae*, anaerobic digestion, combustion. *Separation stage.* Distillation, molecular sieves	Bioethanol, biogas and electricity	The rice husk biorefinery showed higher yields of biogas and electricity compared to the plantain pseudostem case. In general terms, the production costs were as follows: bioethanol (0.48 USD/kg against 0.82 USD/kg), biogas (0.27 USD/m^3 against 0.56 USD/m^3) and electricity (0.02 USD/kW against 0.03 USD/kW)	[33]

(Continued)

TABLE 2.2 (*Continued*)

Examples of Bioprocesses and Biorefineries Considering 2G Feedstocks (Based on the Literature)

Feedstock	Type of Process	Platforms	Technologies	Products	Remarks	Reference
Pinus patula and coffee cut-stems	Stand-alone bioprocess and 2G biorefinery	C5/C6 sugars and lignin	*Pretreatment stage.* Chipping, drying, acid hydrolysis. *Reaction stage.* Gasification, enzymatic hydrolysis, fermentation using *Z. mobilis* and *Thermoanaerobacterium thermosaccharolyticum*. *Separation stage.* Gas turbine, distillation, molecular sieves, membranes	Hydrogen, electricity and ethanol	Some scenarios were assessed from techno-economic, energy, and environmental points of view. The results showed that the hydrogen production based on the biorefinery concept can improve the energy efficiency and profitability, and reduce the emissions of the processes compared to the stand-alone bioprocess.	[34,35]
Olive stone	2G biorefinery	C5/C6 sugars and lignin	*Pretreatment stage.* Acid hydrolysis. *Reaction stage.* Detoxification, enzymatic hydrolysis, fermentation using *Z. mobilis*, *Candida eutropha*, and *Candida mogii*, dehydration reaction and gasification. *Separation stage.* Evaporation, crystallization, distillation, molecular sieves, decantation, gas turbine	Xylitol, furfural, ethanol, PHB, cogeneration system	Two biorefineries were evaluated taking into account the techno-economic and environmental aspects. The first biorefinery scheme described the integrated production of xylitol, furfural, ethanol, and poly-3-hydroxybutyrate (PHB), whereas the second biorefinery scheme considered these products integrated to a cogeneration system. The biorefinery that did not consider the cogeneration system presented an economic margin of 53%, whereas the second biorefinery evidenced an economic margin of 6%.	[36]

(Continued)

TABLE 2.2 (*Continued*)

Examples of Bioprocesses and Biorefineries Considering 2G Feedstocks (Based on the Literature)

Feedstock	Type of Process	Platforms	Technologies	Products	Remarks	Reference
Wood residue (*P. patula* bark)	2G biorefinery	C5/C6 sugars and lignin	*Pretreatment stage.* Milling, acid and alkali hydrolysis, liquid hot water *Reaction stage.* Enzymatic hydrolysis, fermentation using *S. cerevisiae*, dehydration reaction using sulfuric acid as a catalyst, neutralization, gasification *Separation reaction.* Distillation and molecular sieves, decantation, gas turbine	Ethanol, furfural, cogeneration system	Three scenarios were assessed from techno-economic and environmental points of view, taking into account different integration levels—scenario 1: nonintegrated, scenario 2: fully energy integrated, and scenario 3: fully integrated and cogeneration system. According to the results, positive net present values were evidenced in the biorefinery that involve different energy integration levels (scenario 3: 184.54 million USD/year). The same behavior was evidenced in the environmental aspect, and the biorefinery is positively affected by increasing the level of energy integration.	[37]
Sugarcane bagasse	2G biorefinery	C6/C5 sugars and lignin	*Pretreatment stage.* Acid hydrolysis *Reaction stage.* Enzymatic hydrolysis, fermentation using *Z. mobilis* and *C. necator*, gasification	Ethanol, PHB, electricity	An optimization procedure was used in order to select the most promising process route taking into account technical, economic, and environmental aspects. Three biorefinery scenarios were considered. Scenario 1 consisted in energy cogeneration, scenario 2 was an arbitrary distribution, and scenario 3 corresponded to the preselected pathway using the optimization subroutine. Scenario 3 had the best results with an economic margin of 58.83% and biological GHG emissions of 1.55 kg of biological CO_2-e/kg of bagasse.	[38]

(Continued)

TABLE 2.2 (Continued)

Examples of Bioprocesses and Biorefineries Considering 2G Feedstocks (Based on the Literature)

Feedstock	Type of Process	Platforms	Technologies	Products	Remarks	Reference
Brewer's spent grains (BSG)	2G biorefinery	C6/C5 sugars and lignin	*Pretreatment stage.* Washing, acid hydrolysis, milling *Reaction stage.* Detoxification, pulping, enzymatic hydrolysis, fermentation using *C. guilliermondii* and *Lactobacillus delbrueckii*; carbonization reaction *Separation stage.* Ultrafiltration, nanofiltration, vacuum filtration, evaporation, crystallization, ion exchange resin column, adsorption, distillation	Xylitol, lactic acid, activated carbon and phenolic acids (*p*-coumaric and ferulic acid)	Four scenarios based on different levels of heat and mass integration were evaluated considering techno-economic and environmental aspects. The configuration that considered the full mass integration (water recycling) and full-energy integration presented the best economic margin (62.25%) and environmental performance (carbon footprint of 0.96 kg CO_2-e/kg of BSG).	[39]

sorghum, barley, and wheat. The advantage of these residues is that they do not need extra soil to be obtained in comparison with regular crops. However, the total elimination of crop residues can affect both soil properties and the environment, thus reducing the crop productivity. Forest residues, for example, consider logging residues generated in harvesting and milling operations of wood. However, not all these residues can be used as feedstocks due to the nutrient and carbon balance that they can represent for the crop. Finally, energy crops are subdivided into perennial forage crops (e.g., switchgrass and miscanthus) and woody energy crops (a general description of dedicated energy crops was presented in Section 2.1) [16].

2.1.3 3G Feedstocks

The 3G feedstocks include mainly microalgae, which produce high amounts of lipids, carbohydrates, and protein per unit of land [16]. This type of feedstock has been considered of interest due to its fast growth rate in comparison with terrestrial plants and its versatility in terms of substrate and nutrient requirements. Microalgae and algae can grow in waste water, saline water, seawater, carbon dioxide, glycerol, and nonarable lands [16,17]. Additionally, 3G sources have remarkable advantages such as low cost in the cultivation, high-energy production, eco-friendly, and entirely renewable [20,40]. Some limitations of microalgae cultivation are low concentrations; considerable high-energy consumption due to the use of techniques such as mixing, filtration, and centrifugation; contamination problems; and low-lipid generation [17]. Table 2.3 shows some examples of bioprocesses and biorefineries using 3G feedstocks or combination of different generations.

2.1.4 4G Feedstocks

The 4G feedstocks include algae [17,18], industrial waste CO_2, and captured and recycled carbon are considered [20]. Some advantages of these feedstocks are good yields, high capacity of CO_2 capture, and high production rate. Actually, the main limitation is related to the immaturity of the technologies [17]. Examples of this type of biorefineries are still difficult to find in the current literature since the main advances are at the stand-alone processes of valorization but not in a biorefinery system.

2.2 Products

According to the International Energy Agency (IEA), biorefineries can be classified based on the type of obtained products into two groups: energy-driven and product-driven biorefineries [11,12,43].

TABLE 2.3

Examples of Bioprocesses and Biorefineries Considering 3G Feedstocks (Based on the Literature)

Feedstock	Type of Process	Platforms	Technologies	Products	Remarks	Reference
Chlorella vulgaris cake	Stand-alone bioprocess	C6 sugar	*Reaction stage.* Enzymatic hydrolysis, fermentation with *S. cerevisiae.* *Separation stage.* Extractive distillation, conventional distillation, molecular sieves	Ethanol	The global yield of ethanol was 211.9 L per ton of cake. Molecular sieves was the most promising technology, and the total production cost was 0.76 USD per liter, which resulted in an economic margin of 19.15%.	[41]
Castor bean seeds and microalgae *Chlorella* sp.	2G and 3G biorefineries	Castor and microalgae oil, lignin	*Reaction stage.* Epoxidation reaction, biomass gasification combined cycle cogeneration, transesterification reaction *Separation stage.* Distillation and vacuum distillation, mechanical (pressing) and solvent extraction.	Polyol, ethylene-glycol, omega-3 acid, biodiesel, methanol, heat and power	Three scenarios based on different levels of mass and energy integration were simulated and assessed from techno-economic and environmental points of view. The scenario with the best economic and environmental performances was the one including full mass integration, full heat integration, and cogeneration scheme.	[19]

(Continued)

TABLE 2.3 (*Continued*)

Examples of Bioprocesses and Biorefineries Considering 3G Feedstocks (Based on the Literature)

Feedstock	Type of Process	Platforms	Technologies	Products	Remarks	Reference
Sugarcane and microalgae *Chlorella* sp.	1G, 2G, and 3G biorefineries	Microalgae oil, C6 sugar, lignin	*Pretreatment stage.* Milling, washer, drying, acid hydrolysis *Reaction stage.* Clarification, fermentation, gasification, transesterification reaction *Separation stage.* Evaporation, crystallization, distillation, molecular sieves, heat steam recovery, gas turbine, reverse osmosis, absorption, cooling tower, decantation	Sugar, ethanol, electricity, biodiesel, glycerol	Two scenarios were evaluated. Scenario 1 considered the production of sugar, ethanol, and electricity. Scenario 2 was an extension of scenario 1 also considering *Chlorella* sp. cultivation using CO_2-rich streams derived from fermentation and cogeneration systems. Scenario 2 had the best performance from environmental and economic perspectives.	[42]

1. *Energy-driven biorefinery.* The main products that are used for energy generation are (i) gaseous biofuels (biogas, syngas, hydrogen, bio-methane), (ii) solid biofuels (pellets, lignin, charcoal), and (iii) liquid biofuels (biodiesel, bioethanol, Fischer-Tropsch fuels, bio-oil), and (iv) electricity and heat.

2. *Product-driven biorefinery.* The products that are not used for energy generation are (i) chemicals (fine chemicals, building blocks), (ii) organic acids, (iii) polymers and resins, (iv) biomaterials, (v) food and animal feed, and (vi) fertilizers.

Some products can be used as energy products and material products. For example, bioethanol and biohydrogen can be used for both classifications. For this reason, the target market should be identified to know the product classification.

A brief description of the main products of biorefineries can be found in Sections 2.2.1–2.2.4.

2.2.1 Biofuels

Bioethanol, biodiesel, biobutanol, biogas, biohydrogen, and jet biofuel are the main biofuels that a biorefinery can consider as products [44–46]. Table 2.4 shows some examples of different processes for biofuel production. Most of these processes are produced from any generation of feedstocks. Bioethanol and biodiesel are the most common biofuels in the world with the highest production levels. According to the U.S. EIA in 2012, 1,901,000 barrels per day of biofuels were produced worldwide, where 22.7% and 77.3% corresponded to biodiesel and bioethanol, respectively. According to the statistics report by Organization for Economic Cooperation and Development—Food and Agriculture Organization (OECD–FAO), the ratio between the use of bio-diesel to bioethanol will remain the same until 2024 [47].

The United States and Brazil are the two largest producers of bioethanol in the world, accounting together for 90% of the global production [48]. Corn and sugarcane are the main feedstocks to produce bioethanol in the United States and Brazil, respectively. Colombia also produced bioethanol using sugarcane juice and molasses as feedstocks occupying the tenth position at the global level in 2010 with a production of 85 million gallons [49].

Biodiesel is obtained through the transesterification process of edible oils obtained from oilseeds crops such as rapeseed, soybean, sunflower oils, and palm oil with low-molecular-weight alcohols such as methanol or ethanol [50]. Ethanol is produced from biomass through the fermentation of starches and sugars using *Saccharomyces cerevisiae* or *Zymomonas mobilis* with different techniques such as separate hydrolysis and fermentation [51,52], simultane-ous saccharification and fermentation [53,54], and simultaneous saccharifica-tion and co-fermentation [55]. The bioethanol production is considered as a

TABLE 2.4

Conversion Routes for Biofuel Production

Biofuel	Conversion Route	Catalyst or Microorganism	Building Block	Reference
Ethanol	Biochemical conversion method; fermentation	*Zymomonas mobilis* or *Saccharomyces cerevisiae*	Sugars	[44]
	Thermochemical and biological conversion methods; gasification and fermentation	*Alkalibaculum bacchi* and *Clostridium propionicum*	Syngas	[56]
	Chemical and biological conversion methods; esterification, transesterification, and fermentation	*Escherichia coli*	Glycerol	[57]
Butanol	Biochemical conversion method; fermentation	*Clostridium beijerinckii* and/ or *C. acetobutylicum*	Sugars	[60]
	Thermochemical and biological conversion methods; gasification and fermentation	*A. bacchi* and *C. propionicum*	Syngas	[56]
	Chemical and biological conversion methods; esterification, transesterification, and fermentation	*Clotridium pasteurianum*	Glycerol	[61]
Hydrogen	Biochemical conversion method; anaerobic fermentation	*Enterobacter aerogenes*	Glycerol Sugars	[69] [70,72]
Jet fuel range alkanes	Chemical conversion methods; hydrolysis, dehydration, aldol condensation, hydrogenation, and hydrodeoxygenation	Ru/Al_2O_3 or $Pd/MgO-ZrO_2$ $Pd/MgO-ZrO_2$	Xylose/furfural Glucose/ hydroxymethylfurfural (HMF)	[73,74] [73]

versatile process because bioethanol can also be produced through the fermentation of syngas and glycerol using *Alkalibaculum bacchi* and *Escherichia coli*, respectively [56,57].

Biobutanol is an alternative to ethanol with better properties and characteristics. However, this biofuel presents some problems with yields at the reaction stage and then at its purification stage due to the low concentration of complex mixtures including azeotropes. Biobutanol has many advantages over bioethanol; one of them is the larger number of carbons that provide a higher energy content to butanol compared with bioethanol. Additionally, the biobutanol is less corrosive, is less volatile, and presents better blending ability [58,59]. This biofuel can be produced through acetone–butanol–ethanol fermentation using *Clostridium* species to ferment sugars from biomass [60]. Syngas and glycerol can also be used as platforms to produce biobutanol [61].

Biogas is defined as a versatile and sustainable energy carrier. Different energy crops can be converted into biogas through a fermentation process known as anaerobic digestion. Depending on its calorific value, it can be used for different purposes such as chemical synthesis, generating heat or electricity [62]. Biogas is extremely efficient in reducing overall CO_2 emissions in comparison with fossil transportation fuels such as petrol and diesel. Germany is considered as a leader country in the biogas production in all European countries accounting for half of the European primary energy output (50.5% in 2009) and half of the biogas source for electricity generation (49.9% in 2009) [63–65].

Biohydrogen has been identified as one of the energy carriers with the highest potential due to its high-energy density and efficient conversion to electricity. Among the biological methods of producing hydrogen, dark fermentation has some advantages over the photosynthetic and photolytic bioprocesses because of the lower net energy input, higher rate, and moderate yields [66]. Many experts believe that hydrogen is a potential major fuel for transportation in the future because of its eco-friendly combustion products. However, its main potential for road transportation lies in its future role in fuel cell vehicles, and thus, it may largely depend on the development of this technology [67,68]. Biohydrogen production through fermentation can use sugars as platforms and *Clostridium* sp. and *Enterobacter aerogenes* as microorganisms [69,70]. However, the biohydrogen production and storage has many technical and economic challenges in the future.

Jet biofuels are a subject of deep research in the past few years with different combinations of the abovementioned biofuels (mainly biodiesel and bioethanol) for laboratory and pilot setups as well as in sample commercial flights. A more specific example of a pathway for jet biofuel production is the conversion of hemicellulose to xylose through acid hydrolysis, and then xylose is dehydrated to produce furfural. Furfural can be a new platform in the catalytic production of furan-based compounds that are a highly versatile and key derivative in the manufacture of a wide range of important chemicals. Additionally, furfural can serve as precursors in the production of jet fuel substitutes or additives (alkanes C_7–C_{15}) [71].

2.2.2 Bioenergy

Biomass is the most used renewable resource in the world, and it has a great energy potential. This resource can be obtained from plants, plant-derived materials, and animals. Wood has been the first type of biomass used for cooking and heating water, and still remains as the largest biomass energy source, but recently other biomasses can also be used to produce energy. These include food crops, grassy and woody plants, residues from agriculture or forestry, oil-rich algae, and the organic components of municipal and industrial wastes [75]. These residues can provide an array of benefits:

- The use of biomass energy has the potential to greatly reduce GHG emissions. Burning biomass releases the same amount of carbon dioxide as burning fossil fuels. However, biomass releases carbon dioxide that is largely balanced by the carbon dioxide captured in plant growth (depending on how much energy was used to grow, harvest, and process the fuel). It is considered by several authors that biomass has zero CO_2 emissions into the atmosphere, but there are several factors that can increase the emissions of these residues such as the energy used to grow, harvest, and consume the fuel in different stages of the crop cultivation.
- The use of biomass can reduce the dependence on fossil fuels; nowadays, biofuels are the only renewable liquid transportation fuels available.

There are different technologies to transform the biomass into energy: thermal processes such as combustion and gasification to obtain heat to produce steam and subsequently electricity; these technologies can also produce directly bioenergy. The transformation technologies and the biomass availability are the current constraints for energy exploitation. The energy efficiency of the technologies for biomass transformation can vary from 20% (e.g., microturbines) to 50% (e.g., gasification) [75].

2.2.3 Bio-based Materials

Bio-based materials are proteins, starch, cellulose, and latex, among others. They have been attracting more and more attention in the past few years [76]. Bio-based materials—including bio-based polymers, lubricants, solvents, and surfactants—are found to be an interesting option due to its emission reduction potential in the short and the long term. Biopolymers can be classified as biomaterials such as cellulose, starch, and polyhydroxyalkanoates (PHAs), which are the first kind of natural occurring biopolymers. The second kind of biopolymers is polylactic acid (PLA) that can be synthesized from biologically obtained lactic acid, or even polyethylene that is produced from bio-ethanol-based ethylene.

Naturally occurring biopolymers are derived from four broad feedstock areas such as, animal sources provide collagen and gelatin, whereas marine sources provide chitin, which is processed into chitosan. Microbial biopolymer feedstocks are able to produce PLA and PHAs [77]. Agricultural feedstocks are the source of hydrocolloids, lipids, and fatty biopolymers. In principle, biodegradable polymers can also be manufactured entirely from petrochemical raw materials. Currently, there is a range of biopolymers with an interesting potential for substitution of the synthetic polymers. Biopolymers, mainly in a biorefinery, can be considered as products of different types, such as polysaccharides (starch and cellulose polymers), polyesters (PLA and PHAs), polyurethanes, and polyamides. Table 2.5 shows some examples of the different pathways to obtain biopolymers.

Starch polymers have been used in packaging and short-lived consumer goods. Today, starch plastics are one of the most important polymers in the bio-based polymer market. In Europe, the production capacity of starch plastics increased from 30,000 metric tons per year in 2003 to approximately 130,000 metric tons in 2007 [76]. Most of the starch plastics are used for packaging applications, including soluble films for industrial packaging, films for bags and sacks, and loose fills. The raw materials for starch plastic are corn (maize), wheat, potato, cassava, tapioca, and rice [78].

The range of raw materials suitable for lactic acid fermentation includes six-carbon sugars such as glucose together with a large number of compounds that can be easily split into hexoses (e.g., sugars, molasses, sugar beet juice, sulfite liquors, and whey, as well as rice, wheat, and potato starches) [79]. Then, the obtained free acid is purified to yield the product quality required for chemical synthesis. Lactic acid is the precursor molecule for PLA, and it is used for a wide range of application areas, such as packaging

TABLE 2.5

Conversion Routes for Biopolymer Production

Biomaterial	Conversion Route	Catalyst	Building Block	Reference
PHB	Biochemical conversion method; fermentation	*Bacillus megaterium*	Sugars Glycerol	[78]
PHAs	Biochemical conversion method; fermentation	*Bacillus* sp. 871 and 112A	Lactose, galactose, fructose, glucose, mannitol, sucrose, maltose, starch	[80]
		Bacillus sp. INT005	Glucose, butyrate, hexanoate, valerate, octanoate, decanoate	
PLA	Biochemical conversion method. Fermentation	*Escherichia coli* (PLA as homopolymer and P(3HB-*co*-LA) as copolymer)	Sugars	[81]

(cups, bottles, films, and trays), textiles (shirts, furniture), nonwovens (diapers), electronics (mobile phone housing), agriculture (usually blended with thermoplastic starch), and cutlery [76].

PHAs constitute a class of bio-based polyesters with highly attractive qualities for thermo-processing applications. They are produced directly via the fermentation of carbon substrate within the microorganism. They accumulate as granules within the cytoplasm of cells and serve as a microbial energy reserve material. Currently, feedstocks for PHA production are high-value substrates such as sucrose, vegetable oils, and fatty acids. In theory, any carbon source can be utilized, including lignocellulosics from agricultural by-products.

Biopolymers are an interesting alternative to the substitution of synthetic polymers. However, the production of biopolymers such as starch polymers, PLA, and PHAs is still under development. Some reports show that their application and massive use must generate a chain of added values in the biorefinery. Additionally, biopolymers have environmental benefits. However, some obstacles (especially in the production cost and material quality) have limited their use. Thus, research and development is the only way for the improvement of the biopolymer production process, starting with the use of alternative and cheaper raw materials and also the separation and purification methods [78].

2.2.4 Natural Products and Biomolecules

Biomolecules have great importance and potential in the market of some types of biomass. These compounds are widely used as ingredients and feedstocks in food, cosmetics, and pharmaceutical applications [82]. Biomolecules can be classified as follows:

- Essential oils are liquids that contain the volatile compounds from plants. They are utilized in the food (flavor), cosmetic (fragrance), and pharmaceutical industries.
- Oleoresins comprise essential oils, resins, and others that form the volatile and nonvolatile fractions of the species from which they are extracted. These are commonly used in the food industry.
- Pigments and natural dyes are used in the color industry.

Bio-based sources are power machines to produce a broad variety of natural products, also known as secondary metabolites [82]. Many of the secondary metabolites exhibit specific biological activities that make them useful in a variety of applications such as antioxidants, antimicrobials, or antitumors [83,84]. Natural products from plants can be classified as nitrogen- and/or sulfur-containing compounds, phenolic compounds, and terpenoids. Phenolic compounds play an important role in biorefineries. Table 2.6 shows

TABLE 2.6

Conversion Routes for Natural Products

Natural Product or Biomolecule	Conversion Route	Feedstock	Reference
Essential oil	Steam distillation	Timber (*Cedrela odorata*)	[82]
Polyphenolic compound extract	Supercritical fluid extraction (SFE)		
Polyphenolic compound extract: seed and peel	SFE	Makambo fruit (*Theobroma bicolor*): pulp, seed, and peel	[25]
Pasteurized pulp	Mechanical pulping and pasteurization		
Cocoa	Mechanical extraction and crushing		
Butter	Mechanical extraction		
Polyphenolic compound extract: peel	SFE	Aguaje fruit (*Mauritia flexuosa*): pulp, seed, and peel	[82]
Pasteurized pulp	Mechanical pulping and pasteurization		
Flour	Mechanical pulping		
Oil	Mechanical extraction		
Antioxidant-rich extract	Conventional solvent extraction (CSE) and SFE	Zapote fruit (*Matisia cordata*)	[85]
		Goldenberry fruit (*Physalis peruviana*)	
		Tamarillo fruit (*Solanum betaceum*)	
		Naiku fruit (*Renealmia alpinia*)	
Pasteurized pulp	Mechanical pulping and pasteurization	Copoazu fruit (*Theobroma grandiflorum*)	[24]
Antioxidant extract	SFE		
Oil seed	Cracking and CSE		
Essential oil	Steam distillation		

some examples of the production pathway of natural products or biomolecules using the biorefinery concept. Here, it is not considered fermentative routes to obtain some of these natural products.

References

1. A. Demirbas, "Chapter 2: Biomass feedstocks," in *Biofuels—Securing the Planet's Future Energy Needs*, London, UK: Springer-Verlag, pp. 45–85, 2009.

2. Office of Energy Efficiency & Renewable Energy, "Bioenergy—Biomass feedstocks," 2017. [Online]. Available: https://energy.gov/eere/bioenergy/biomass-feedstocks (Accessed 01 Aug. 2017.)
3. A. A. Koutinas, A. Vlysidis, D. Pleissner, N. Kopsahelis, I. Lopez Garcia, I. K. Kookos, S. Papanikolaou, T. H. Kwan, C. S. Lin, "Valorization of industrial waste and by-product streams via fermentation for the production of chemicals and biopolymers," *Chem. Soc. Rev.*, vol. 43, no. 8, pp. 2587–2627, 2014.
4. J. A. López, V. Bucalá, and M. A. Villar, "Application of dynamic optimization techniques for poly(β-hydroxybutyrate) production in a fed-batch bioreactor," *Ind. Eng. Chem. Res.*, vol. 49, no. 4, pp. 1762–1769, 2010.
5. M. Laguerre, M. C. Figueroa-Espinoza, M. Pina, M. Benaissa, A. Combe, A. Rossignol Castera, J. Lecomte, P. Villeneuve, "Characterization of olive-leaf phenolics by ESI-MS and evaluation of their antioxidant capacities by the CAT assay," *JAOCS, J. Am. Oil Chem. Soc.*, vol. 86, no. 12, pp. 1215–1225, 2009.
6. F. C. Ogbo, "Conversion of cassava wastes for biofertilizer production using phosphate solubilizing fungi," *Bioresour. Technol.*, vol. 101, pp. 4120–4124, 2010.
7. S. V. Vassilev, D. Baxter, L. K. Andersen, and C. G. Vassileva, "An overview of the chemical composition of biomass," *Fuel*, vol. 89, no. 5, pp. 913–933, May 2010.
8. Independent Statistics & Analysis U.S. Energy Information Administration (EIA), "Glossary," 2017. [Online]. Available: www.eia.gov/tools/glossary/index.php?id=Biomass (Accessed 01 Aug. 2017.)
9. J. G. Speight, "Chapter 8: Fuels from biomass," in *Synthetic Fuels Handbook: Properties, Process, and Performance*, New York: The McGraw-Hill Companies, Inc., 2008, pp. 221–230.
10. W. Xiaohua and F. Zhenmin, "Biofuel use and its emission of noxious gases in rural China," *Renew. Sustain. Energy Rev.*, vol. 8, no. 2, pp. 183–192, Apr. 2004.
11. IEA Bioenergy Task42, "Biorefineries: Adding value to the sustainable utilisation of biomass," IEA Bioenergy, vol. 1, pp. 1–16, 2009.
12. F. Cherubini et al., "Toward a common classification approach for biorefinery systems," *Biofuels, Bioprod. Bior.*, vol. 3, pp. 534–546, 2009.
13. F. Cherubini, "The biorefinery concept: Using biomass instead of oil for producing energy and chemicals," *Energy Convers. Manag.*, vol. 51, no. 7, pp. 1412–1421, 2010.
14. B. Kamm and M. Kamm, "Principles of biorefineries," *Appl. Microbiol. Biotechnol.*, vol. 64, pp. 137–145, 2004.
15. F. Cherubini and S. Ulgiati, "Crop residues as raw materials for biorefinery systems—A LCA case study," *Appl. Energy*, vol. 87, pp. 47–57, 2010.
16. M. A. Carriquiry, X. Du, and G. R. Timilsina, "Second generation biofuels: Economics and policies," *Energy Policy*, vol. 39, no. 7, pp. 4222–4234, 2011.
17. K. Dutta, A. Daverey, and J. G. Lin, "Evolution retrospective for alternative fuels: First to fourth generation," *Renew. Energy*, vol. 69, pp. 114–122, 2014.
18. J. Lü, C. Sheahan, and P. Fu, "Metabolic engineering of algae for fourth generation biofuels production," *Energy Environ. Sci.*, vol. 4, pp. 2451–2466, 2011.
19. J. Moncada, C. A. Cardona, and L. E. Rincón, "Design and analysis of a second and third generation biorefinery: The case of castorbean and microalgae," *Bioresour. Technol.*, vol. 198, pp. 836–843, Dec. 2015.
20. A. K. Azad, M. G. Rasul, M. M. K. Khan, S. C. Sharma, and M. A. Hazrat, "Prospect of biofuels as an alternative transport fuel in Australia," *Renew. Sustain. Energy Rev.*, vol. 43, pp. 331–351, Mar. 2015.

21. S. N. Naik, V. V. Goud, P. K. Rout, and A. K. Dalai, "Production of first and second generation biofuels: A comprehensive review," *Renew. Sustain. Energy Rev.*, vol. 14, pp. 578–597, Feb. 2010.

22. A. Mohr and S. Raman, "Lessons from first generation biofuels and implications for the sustainability appraisal of second generation biofuels," *Energy Policy*, vol. 63, pp. 114–122, 2013.

23. R. E. H. Sims, W. Mabee, J. N. Saddler, and M. Taylor, "An overview of second generation biofuel technologies," *Bioresour. Technol.*, vol. 101, pp. 1570–1580, Mar. 2010.

24. I. X. Cerón, J. C. Higuita, and C. A. Cardona, "Analysis of a biorefinery based on *Theobroma grandiflorum* (copoazu) fruit," *Biomass Convers. Biorefinery*, vol. 5, pp. 183–194, 2015.

25. A. A. González, J. Moncada, A. Idarraga, M. Rosenberg, and C. A. Cardona, "Potential of the Amazonian exotic fruit for biorefineries: The *Theobroma bicolor* (Makambo) case," *Ind. Crops Prod.*, vol. 86, pp. 58–67, 2016.

26. J. M. Naranjo, C. A. Cardona, and J. C. Higuita, "Use of residual banana for polyhydroxybutyrate (PHB) production: Case of study in an integrated biorefinery," *Waste Manag.*, vol. 34, pp. 2634–2640, 2014.

27. L. E. Rincón, J. Moncada, and C. A. Cardona, "Analysis of potential technological schemes for the development of oil palm industry in Colombia: A biorefinery point of view," *Ind. Crops Prod.*, vol. 52, pp. 457–465, Jan. 2014.

28. Ó. J. Sánchez and C. A. Cardona, "Conceptual design of cost-effective and environmentally-friendly configurations for fuel ethanol production from sugarcane by knowledge-based process synthesis," *Bioresour. Technol.*, vol. 104, pp. 305–314, 2012.

29. J. Moncada, M. M. El-Halwagi, and C. A. Cardona, "Techno-economic analysis for a sugarcane biorefinery: Colombian case," *Bioresour. Technol.*, vol. 135, pp. 533–543, 2013.

30. J. Moncada, J. Tamayo, and C. A. Cardona, "Evolution from biofuels to integrated biorefineries: Techno-economic and environmental assessment of oil palm in Colombia," *J. Clean. Prod.*, vol. 81, pp. 51–59, 2014.

31. J. A. Dávila, M. Rosenberg, and C. A. Cardona, "Techno-economic and environmental assessment of p-cymene and pectin production from orange peel," *Waste and Biomass Valorization*, vol. 6, pp. 253–261, 2015.

32. J. A. Dávila, M. Rosenberg, and C. A. Cardona, "A biorefinery for efficient processing and utilization of spent pulp of Colombian Andes Berry (*Rubus glaucus* Benth.): Experimental, techno-economic and environmental assessment," *Bioresour. Technol.*, vol. 223, pp. 227–236, 2017.

33. L. V. Daza Serna, J. C. Solarte Toro, S. Serna Loaiza, Y. Chacón Perez, and C. A. Cardona Alzate, "Agricultural waste management through energy producing biorefineries: The Colombian case," *Waste Biomass Valori.*, vol. 7, pp. 789–798, 2016.

34. C. A. García, R. Betancourt, and C. A. Cardona, "Stand-alone and biorefinery pathways to produce hydrogen through gasification and dark fermentation using *Pinus Patula*," *J. Environ. Manage.*, vol. 203, pp. 695–703, 2016.

35. C. A. García, J. Moncada, V. Aristizábal, and C. A. Cardona, "Techno-economic and energetic assessment of hydrogen production through gasification in the Colombian context: Coffee Cut-Stems case," *Int. J. Hydrogen Energy*, vol. 42, pp. 5849–5864, 2017.

36. V. Hernández, J. M. Romero-García, J. A. Dávila, E. Castro, and C. A. Cardona, "Techno-economic and environmental assessment of an olive stone based biorefinery," *Resour. Conserv. Recycl.*, vol. 92, pp. 145–150, 2014.

37. J. Moncada, C. A. Cardona, J. C. Higuita, J. J. Vélez, and F. E. López-Suarez, "Wood residue (*Pinus patula* bark) as an alternative feedstock for producing ethanol and furfural in Colombia: experimental, techno-economic and environmental assessments," *Chem. Eng. Sci.*, vol. 140, pp. 309–318, Feb. 2016.

38. J. Moncada, L. G. Matallana, and C. A. Cardona, "Selection of process pathways for biorefinery design using optimization tools: A Colombian case for conversion of sugarcane bagasse to ethanol, poly-3-hydroxybutyrate (PHB), and energy," *Ind. Eng. Chem. Res.*, vol. 52, no. 11, pp. 4132–4145, 2013.

39. S. I. Mussatto, J. Moncada, I. C. Roberto, and C. A. Cardona, "Techno-economic analysis for brewer's spent grains use on a biorefinery concept: The Brazilian case," *Bioresour. Technol.*, vol. 148, pp. 302–310, 2013.

40. L. Brennan and P. Owende, "Biofuels from microalgae—A review of technologies for production, processing, and extractions of biofuels and co-products," *Renew. Sustain. Energy Rev.*, vol. 14, pp. 557–577, 2010.

41. J. Moncada, J. J. Jaramillo, J. C. Higuita, C. Younes, and C. A. Cardona, "Production of bioethanol using *Chlorella vulgaris* cake: A technoeconomic and environmental assessment in the Colombian context," *Ind. Eng. Chem. Res.*, vol. 52, pp. 16786–16794, 2013.

42. J. Moncada, J. A. Tamayo, and C. A. Cardona, "Integrating first, second, and third generation biorefineries: Incorporating microalgae into the sugarcane biorefinery," *Chem. Eng. Sci.*, vol. 118, pp. 126–140, 2014.

43. R. van Ree and A. van Zeeland, "IEA Bioenergy—Task42 Biorefining," 2014.

44. V. Aristizábal, Á. Gómez, and C. A. Cardona, "Biorefineries based on coffee cut-stems and sugarcane bagasse: Furan-based compounds and alkanes as interesting products," *Bioresour. Technol.*, vol. 196, pp. 480–489, 2015.

45. V. Aristizábal, C. A. García, and C. A. Cardona, "Integrated production of different types of bioenergy from oil palm through biorefinery concept," *Waste Biomass Valori.*, vol. 7, no. 4, pp. 737–745, 2016.

46. M. Demuez, A. Mahdy, E. Tomás-Pejó, C. González-Fernández, and M. Ballesteros, "Enzymatic cell disruption of microalgae biomass in biorefinery processes," *Biotechnol. Bioeng.*, vol. 112, pp. 1955–1966, 2015.

47. Food and Agriculture Organization of The United Nations (FAO), "OCDE-FAO perspectivas agrícolas 2015-2024," 2015. [Online]. Available: http://www.fao.org/3/a-i4738s.pdf , (Accessed Aug. 2017.).

48. G. Koçar and N. Civaş, "An overview of biofuels from energy crops: Current status and future prospects," *Renew. Sustain. Energy Rev.*, vol. 28, pp. 900–916, Dec. 2013.

49. P. A. Cremonez et al., "Current scenario and prospects of use of liquid biofuels in South America," *Renew. Sustain. Energy Rev.*, vol. 43, pp. 352–362, 2015.

50. L. F. Gutiérrez, O. J. Sánchez, and C. A. Cardona, "Process integration possibilities for biodiesel production from palm oil using ethanol obtained from lignocellulosic residues of oil palm industry," *Bioresour. Technol.*, vol. 100, no. 3, pp. 1227–1237, Feb. 2009.

51. J. A. Quintero, J. Moncada, and C. A. Cardona, "Techno-economic analysis of bioethanol production from lignocellulosic residues in Colombia: A process simulation approach," *Bioresour. Technol.*, vol. 139, pp. 300–307, Jul. 2013.

52. K. J. Dussán, D. D. V. Silva, V. H. Perez, and S. S. da Silva, "Evaluation of oxygen availability on ethanol production from sugarcane bagasse hydrolysate in a batch bioreactor using two strains of xylose-fermenting yeast," *Renew. Energy*, vol. 87, pp. 703–710, 2016.

53. A. Gladis, P.-M. Bondesson, M. Galbe, and G. Zacchi, "Influence of different SSF conditions on ethanol production from corn stover at high solids loadings," *Energy Sci. Eng.*, vol. 3, no. 5, pp. 481–489, 2015.

54. F. A. Gonçalves, H. A. Ruiz, C. D. C. Nogueira, E. S. Dos Santos, J. A. Teixeira, and G. R. De Macedo, "Comparison of delignified coconuts waste and cactus for fuel-ethanol production by the simultaneous and semi-simultaneous saccharification and fermentation strategies," *Fuel*, vol. 131, pp. 66–76, 2014.

55. M. Yasuda, H. Nagai, K. Takeo, Y. Ishii, and K. Ohta, "Bio-ethanol production through simultaneous saccharification and co-fermentation (SSCF) of a low-moisture anhydrous ammonia (LMAA)-pretreated napiegrass (*Pennisetum purpureum* Schumach)," *Springerplus*, vol. 3, no. 1, p. 333, 2014.

56. K. Liu, H. K. Atiyeh, B. S. Stevenson, R. S. Tanner, M. R. Wilkins, and R. L. Huhnke, "Continuous syngas fermentation for the production of ethanol, n-propanol and n-butanol," *Bioresour. Technol.*, vol. 151, pp. 69–77, 2014.

57. N. Chaudhary, M. O. Ngadi, B. K. Simpson, and L. S. Kassama, "Biosynthesis of ethanol and hydrogen by glycerol fermentation using *Escherichia coli*," *Adv. Chem. Eng. Sci.*, vol. 1, pp. 83–89, 2011.

58. P. S. Nigam and A. Singh, "Production of liquid biofuels from renewable resources," *Prog. Energy Combust. Sci.*, vol. 37, no. 1, pp. 52–68, 2011.

59. A. Morone and R. A. Pandey, "Lignocellulosic biobutanol production: Gridlocks and potential remedies," *Renew. Sustain. Energy Rev.*, vol. 37, pp. 21–35, Sep. 2014.

60. B. Ndaba, I. Chiyanzu, and S. Marx, "n-Butanol derived from biochemical and chemical routes: A review," *Biotechnol. Reports*, vol. 8, pp. 1–9, 2015.

61. Y. S. Jang, A. Malaviya, C. Cho, J. Lee, and S. Y. Lee, "Butanol production from renewable biomass by Clostridia," *Bioresour. Technol.*, vol. 123, pp. 653–663, 2012.

62. J. Rezaiyan and N. P. Cheremisinoff, *Gasification Technologies : A Primer for Engineers and Scientists*, Boca Raton, FL: CRC Press/Taylor & Francis, 2005.

63. J. Havukainen, V. Uusitalo, A. Niskanen, V. Kapustina, and M. Horttanainen, "Evaluation of methods for estimating energy performance of biogas production," *Renew. Energy*, vol. 66, pp. 232–240, Jun. 2014.

64. E. B. Association, "Biogas Report 2011."

65. E. B. A. (AEBIOM), "A biogas road map for Europe."

66. V. Gadhamshetty, Y. Arudchelvam, N. Nirmalakhandan, and D. C. Johnson, "Modeling dark fermentation for biohydrogen production: ADM1-based model vs. Gompertz model," *Int. J. Hydrog. Energy*, vol. 35, no. 2, pp. 479–490, Jan. 2010.

67. J. Huang and C. H. Wong, "Hydrogen transportation properties in carbon nano-scroll investigated by using molecular dynamics simulations," *Comput. Mater. Sci.*, vol. 102, pp. 7–13, May 2015.

68. B. Saynor and M. Leach, "The potential for renewable energy sources in aviation. PRESAV final report," 2003.

69. M. L. Chong, V. Sabaratnam, Y. Shirai, and M. A. Hassan, "Biohydrogen production from biomass and industrial wastes by dark fermentation," *Int. J. Hydrogen Energy*, vol. 34, no. 8, pp. 3277–3287, 2009.

70. S. Tanisho and Y. Ishiwata, "Continuous hydrogen production from molasses by the bacterium *Enterobacter aerogenes*," *Int. J. Hydrogen Energy*, vol. 19, no. 10, pp. 807–812, 1994.

71. J. N. Chheda and J. A. Dumesic, "An overview of dehydration, aldol-condensation and hydrogenation processes for production of liquid alkanes from biomass-derived carbohydrates," *Catal. Today*, vol. 123, pp. 59–70, May 2007.

72. N. Ren, J. Li, B. Li, Y. Wang, and S. Liu, "Biohydrogen production from molasses by anaerobic fermentation with a pilot-scale bioreactor system," *Int. J. Hydrogen Energy*, vol. 31, no. 15, pp. 2147–2157, 2006.

73. C. J. Barrett, J. N. Chheda, G. W. Huber, and J. A. Dumesic, "Single-reactor process for sequential aldol-condensation and hydrogenation of biomass-derived compounds in water," *Appl. Catal. B Environ.*, vol. 66, no. 1–2, pp. 111–118, 2006.

74. J. Q. Bond et al., "Production of renewable jet fuel range alkanes and commodity chemicals from integrated catalytic processing of biomass," *Energy Environ. Sci.*, vol. 7, no. 4, pp. 1500–1523, 2014.

75. C. A. García Velásquez, "Hydrogen production through gasification and dark fermentation," Master Thesis. Universidad Nacional de Colombia. Departamento de Ingeniería Química, 2016.

76. D. Phylipsen, M. Kerssemeeckers, M. Patel, J. de Beer, and P. Eder, "Clean technologies in the materials sector—Current and future environmental performance of material technologies. European Commission—Institute for Prospective Technological Studies (IPTS)," 2002.

77. J. G. C. Gomez, B. S. Méndez, P. I. Nikel, M. J. Pettinari, M. A. P. And, and L. F. Silva, "Chapter 3: Making green polymers even greener: Towards sustainable production of polyhydroxyalkanoates from agroindustrial by-products," in *Advances in Applied Biotechnology*, Prof. M. Petre Eds., pp. 41–62, Rijeka, Croatia : InTech, 2012.

78. J. M. Naranjo, "Design and analysis of the polyhydroxybutyrate (PHB) production from agroindustrial wastes in Colombia," PhD Thesis. Universidad Nacional de Colombia. Departamento de Ingeniería Eléctrica, Electrónica y Computación, 2014.

79. J. Venus, S. Fiore, F. Demichelis, and D. Pleissner, "Centralized and decentralized utilization of organic residues for lactic acid production," *J. Clean. Prod.*, vol. 172, pp. 778–785, 2018.

80. J. A. López, J. M. Naranjo, J. C. Higuita, M. A. Cubitto, C. A. Cardona, and M. A. Villar, "Biosynthesis of PHB from a new isolated *Bacillus megaterium* strain: Outlook on future developments with endospore forming bacteria," *Biotechnol. Bioprocess Eng.*, vol. 17, no. 2, pp. 250–258, 2012.

81. Y. K. Jung, T. Y. Kim, S. J. Park, and S. Y. Lee, "Metabolic engineering of *Escherichia coli* for the production of polylactic acid and its copolymers," *Biotechnol. Bioeng.*, vol. 105, no. 1, pp. 161–171, 2010.

82. A. A. González, "Design and assessment of high technology processes for enhancing the viability of agribusiness based on the sustainable use of biomass in Amazonas," Universidad Nacional de Colombia. Departamento de Ingeniería Eléctrica, Electrónica y Computación., 2015.

83. J. F. Ayala-Zavala et al., "Agro-industrial potential of exotic fruit byproducts as a source of food additives," *Food Res. Int.*, vol. 44, no. 7, pp. 1866–1874, 2011.

84. G. M. Cragg and D. J. Newman, "Plants as a source of anti-cancer agents," *J. Ethnopharmacol.*, vol. 100, pp. 72–79, 2005.

85. I. X. Cerón Salazar, "Design and evaluation of processes to obtain antioxidant-rich extracts from tropical fruits cultivated in Amazon, Caldas and Northern Tolima regions," PhD Thesis. Universidad Nacional de Colombia. Departamento de Ingeniería Eléctrica, Electrónica y Computación, 2013.

3

Biorefinery Design Strategy: From Process Synthesis to Sustainable Design

This chapter aims to present a heuristic approach related with the design of a biorefinery system using the hierarchy, sequencing, and integration concepts. These concepts are presented as a conceptual tool that allows proposing the better route to convert different types of feedstocks into the desired products through the application of current technologies and strategies for its optimal implementation. Moreover, the hierarchy, sequencing, and integration concepts are explained and discussed using examples of biorefineries reported in the literature aiming to demonstrate the applicability of the design approach.

The conceptual design promotes an approach based on an inclusive design of biorefineries, where new challenges are assumed such as the inclusion of a wide range of feedstocks and the development of local and regional areas. As this methodology includes many factors, it is possible to say that it is based on a holistic approach. This strategy has been applied widely in the process engineering for the design of chemical plants but not for the design of biorefineries [1,2]. Characteristics such as, availability, costs, and market needs are the basis for choosing the feedstocks and products. However, there are some constraints linked to the uses and applications. For example, the value-added products such as antioxidants have priority in their production over energy products such as biofuels.

Three types of analysis are included in the conceptual design of a biorefinery: (i) technical, (ii) economic, and (iii) environmental analyses. The commercial software package Aspen Plus is used as a tool to generate the technical analysis based on mass and energy balances. Then, Aspen Process Economic Analyzer is used to calculate the economic results. Finally, the U.S. Environmental Protection Agency (EPA) Waste Reduction (WAR) algorithm and/or GREENSCOPE or life cycle assessment approaches allow determining the generated environmental impact. Two types of integration can be developed: (i) mass using Aspen Plus and (ii) energy using pinch or energy analysis. Experimental data such as feedstock characterization, reaction kinetics, yields, and operation conditions reported in literature are fed into the simulation tools. A comparison between the results for different scenarios can be made in order to choose those presenting the best performance. The conceptual design applied to biorefineries comes from the synthesis of the chemical process approach [1], [2]. However, the approach is extended to complex systems (i.e., biorefineries with a multiproduct portfolio) because new elements are included. The conceptual design is composed of three concepts: (i) hierarchy, (ii) sequencing, and (iii) integration [1,3,4]. These concepts are interlinked, and their combination allows a process synthesis approach to obtain global biorefinery models. Figure 3.1 shows a

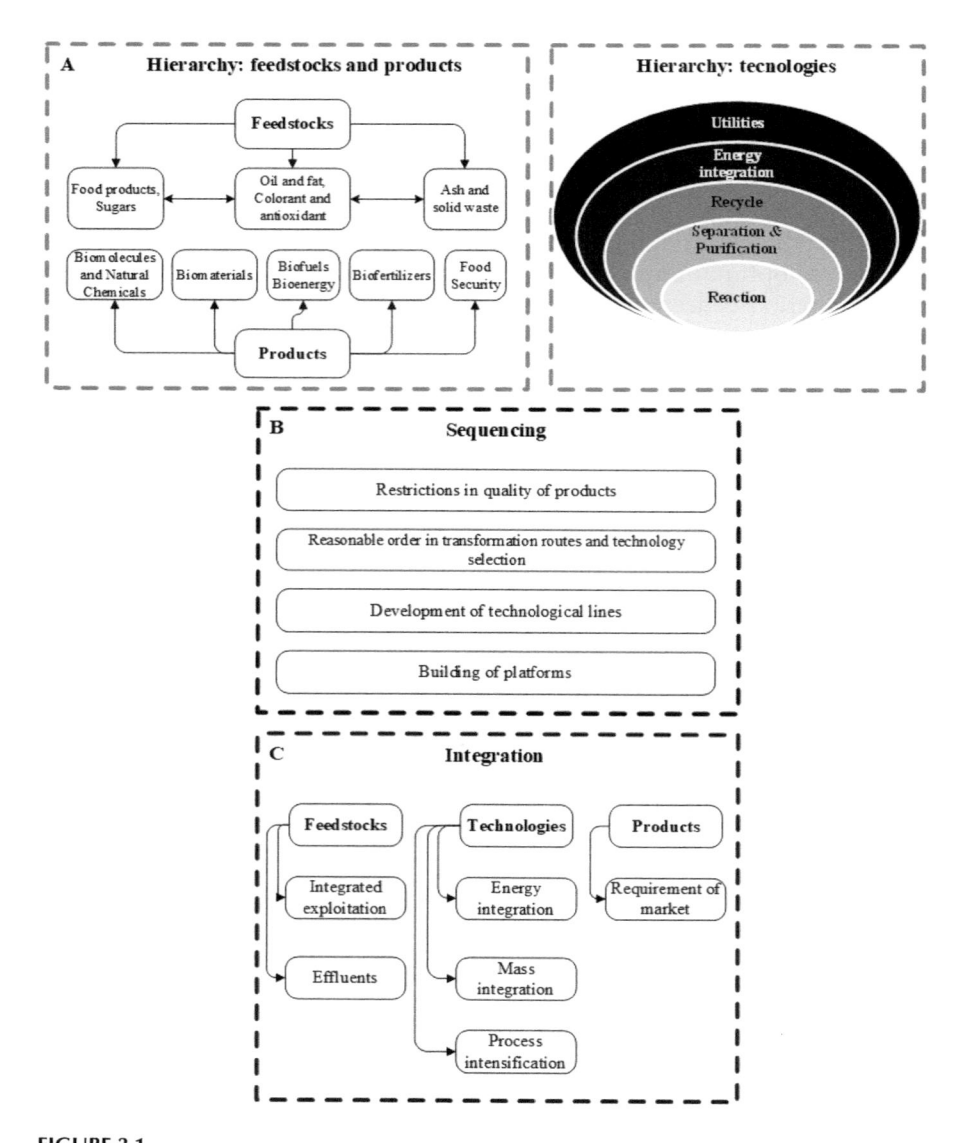

FIGURE 3.1
Proposed design strategy considering the concepts of (A) hierarchy, (B) sequencing, and (C) integration. (Taken from [3].)

graphic description of this strategy. A brief explanation of these concepts is presented in Sections 3.1–3.3.

3.1 Hierarchy Concept

The hierarchical decomposition of feedstocks and products is the first level that must be considered in the design of biorefineries (see Figure 3.1A). Both raw materials and products are related through the feedstock generation and the different platforms to obtain a wide variety of products. In this way, as discussed in Chapter 2, feedstocks are classified into edible crops (first generation); agroindustrial residues, solid waste, residues, and oils and fats (second generation); algae (third generation); and carbon dioxide (fourth generation). On the other hand, products are classified into six main families: biofuels (referring to liquid fuels), bioenergy, biomolecules and natural chemicals, biomaterials, food products, and biofertilizers [5]. Then, the first level of the hierarchical decomposition involves an overall understanding of the feedstock source and the final product applications, favoring the obtainment of high value-added products over the production of bulk chemicals and energy carriers. Therefore, a biorefinery should not be seen only as the counterpart of petrochemical facilities because the final use of the obtained products gives a series of conditions that allow increasing the overall impact of the biorefinery. Thus, different industrial sectors such as pharmaceutical, food, and cosmetic, among others, can be added to the potential markets in which biorefineries can be introduced.

The purpose of a biorefinery system (in its ideal case) is to integrate all feedstock generations as well as all desired products, avoiding as much as possible the production of compounds to be considered finally as residues [5]. This can be achieved only by introducing a well-established conversion pathway that makes use of the best available technologies in the industrial sector. In this sense, the hierarchical decomposition of the technologies is the second level that must be considered in the design of biorefineries. This level is focused on the steps that affect each process or the global biorefinery the most. This is an extension of the well-known onion diagram (see Figure 3.1A), because the reaction step is not always the heart of the process, as it is the situation of most of petrochemical processes [2]. In some cases, the pretreatments are critical to obtain the final products and can be considered as the heart of the design over other processes such as fermentation. For instance, in a lignocellulosic-based biorefinery, the pretreatment and hydrolysis stages are critical to build a sugar-based platform if it is the desired one. Nonetheless, if the desired platform is of thermochemical nature (e.g., syngas, bio-oils), then the critical stage will be a thermochemical process (e.g., gasification, pyrolysis). This hierarchical decomposition is critical because it affects directly the sequence concept interlinking each other [6].

3.1.1 Feedstocks and Products—Hierarchical Decomposition

The first level of the hierarchical analysis (i.e., feedstocks and products) can be observed in the case study presented by Moncada et al. (2014) [7], in which the production of biodiesel, ethanol, and poly-3-hydroxybutyrate (PHB) is performed using as raw materials first and second generation feedstocks such as crude palm oil (CPO) and lignocellulosic biomass (i.e., empty fruit bunches and pressed palm fiber). In this biorefinery, the hierarchical decomposition of the feedstocks and products was performed depending on the type of product to be obtained. In fact, the CPO was used to produce biodiesel based on the large amount of triglycerides that can be transesterified to produce methyl esters. On the other hand, the lignocellulosic biomass was transformed using a biotechnological pathway (i.e., fermentation) to produce fuel ethanol. In addition, this biorefinery shows the use of glycerol to produce PHB, which also demonstrates the applicability of the integration of feedstocks and products to minimize the amount of solid or liquid wastes, increasing the number of value-added products. The biorefinery scheme can be found in Section 7.3.

The hierarchical decomposition of the feedstock was done based on the chemical composition of the raw materials. Thus, a comprehensive analysis of the main elements that compose the feedstock is necessary to visualize the possible products and processes that can be obtained and used, respectively. In this case study, the biorefinery has 1 ton per hour of CPO mass flow of and 0.9 ton per hour of empty fruit bunches with a moisture content of 50%. The mass flows of the biorefinery products were 1,007, 150.20, and 26.20 kg/h of biodiesel, fuel ethanol, and PHB, respectively. These results show that the mass yield of the biodiesel production from CPO is about 1 kg biodiesel/kg of CPO, 211.52 L ethanol/ton empty fruit bunch, and 0.30 kg PHB/kg of glycerol. Therefore, these yields show that a correct hierarchical analysis of the feedstocks and products of a biorefinery system can provide high mass yields and large amounts of products, which can be positively reflected in economic indicators reported by the authors.

Other examples of biorefineries where the hierarchical decomposition of the raw materials was used as a main decision factor to propose the products to be obtained have been reported in literature [6]. These cases are different from the abovementioned ones due to the fact that the feedstock employed in the biorefineries is only a lignocellulosic material, which requires a strategy to select the possible products that can be obtained in function of the chemical composition of the raw material and the possible platforms that can be considered. This is the case of the biorefinery scheme proposed by Dávila et al. (2016) [8], where the ethanol, xylitol, and PHB production is proposed using only the brewer's spent grain as raw material. This feedstock has a high amount of glucose and xylose (i.e., 9.40% and 15.80% w/w dry basis, respectively), which results in the formation of ethanol and xylitol as interesting products. In this biorefinery, the pretreatment stage can be considered as a fractionation stage due to the cellulose and hemicellulose separation from the lignocellulosic matrix of the raw material to be processed in different lines as shown in Figure 3.2. The fraction of cellulose is used to produce glucose as a building block, and the fraction of hemicellulose is used to produce xylose as a platform. However, each of these building blocks is used to produce one product, ethanol and xylitol, respectively. Moreover, the stillage produced in the ethanol productive process is used as a raw material to produce PHB, which minimizes the amount of liquid waste that needs to be treated in a residual water treatment plant.

After the fractionation of the raw material and the ethanol and xylitol production, the biorefinery scheme proposed by the authors suggest the use of the liquid and solid residues from these processing lines to produce PHB, heat, and power through a stillage fermentation and cogeneration system. Then, it is also possible to see in this biorefinery the integration of feedstocks and products aiming to minimize the amount of solid or liquid waste generated while obtaining the main products. The mass yields for ethanol, xylitol, and PHB reported for this case study were 41.5 L/ton of brewer's spent grain, 101.53 kg/ton of brewer's spent grain, and 10.53 kg/ton of brewer's spent grain for ethanol, xylitol, and PHB, respectively. The ethanol yield

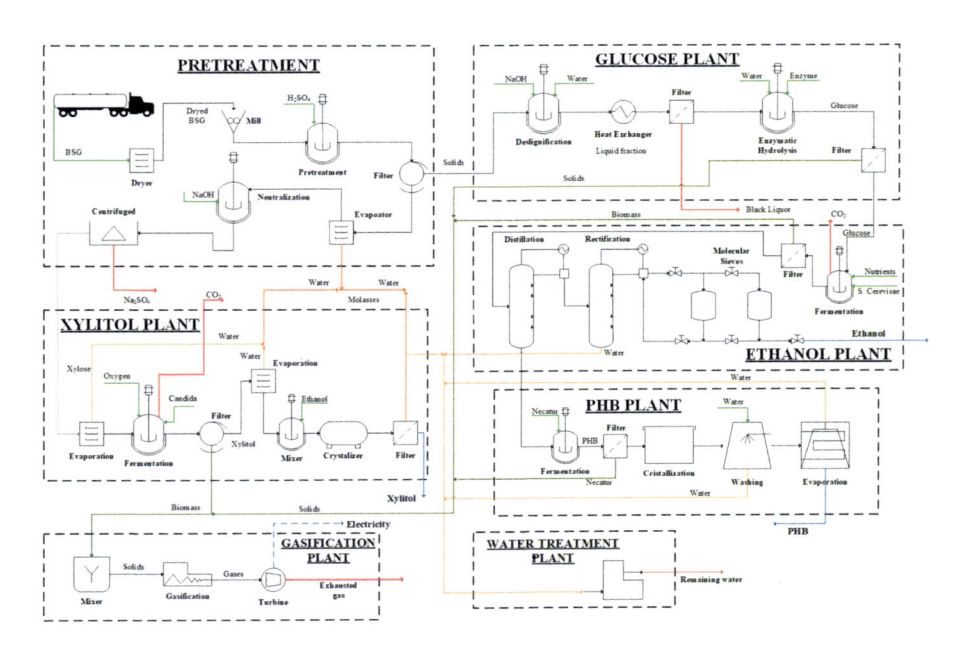

FIGURE 3.2
A schematic diagram of a biorefinery for production of xylitol, ethanol, and PHB from Brewer's spent grain.

reported in this case is lower than the yield reported for the first biorefinery that was analyzed. This difference can be explained due to the cellulose content of both raw materials. The cellulose content in the empty fruit bunches is higher than the brewer's spent grain. In addition, in the first case study, both cellulose and hemicellulose were used as original platforms to produce ethanol, which increased the yield. In addition, the PHB yield reported in the second biorefinery is higher than in the first biorefinery, which suggest that the use of the stillage from the fuel ethanol production is a better substrate to produce PHB than glycerol. Therefore, an integration of residues in the first case study to increase the PHB production may be proposed. As it can be observed in the abovementioned biorefinery cases, the hierarchical decomposition of the feedstocks and products plays an important role during the definition of the biorefinery target. Then, the use of the first-, second-, and third-generation raw materials to produce a wide variety of products is the first step when a biorefinery system is designed.

3.1.2 Technologies—Hierarchical Decomposition

The second level of the hierarchical analysis (i.e., technologies) takes into account a set of available technologies to use in the biorefinery system once the feedstocks and products have been defined. This analysis also was applied in the abovementioned two study cases. However, to complement the information and to give an overall explanation of this concept, other two examples will be discussed in the following text.

The hierarchical selection of the technologies that may be used in a biorefinery is directly associated with the economic and environmental indicators calculated for the biorefinery as well as the mass and energy yields obtained. However, the reactions involved in the desired processing lines are the real basis for reactors and separators. In this way, the biodiesel production using different feedstocks and technologies was studied by Rincón et al. (2014) [9]. In this work, two types of processes to produce biodiesel were assessed. The first one was the biodiesel production using an alkali-catalyzed process and the second one was related with the use of an acid-catalyzed process. Figures 3.3 and 3.4 show the process flow diagrams. According to the authors, the mass yields of the produced biodiesel were higher in the alkali-catalyzed process than in the acid-catalyzed process. The results showed that 1.010 kg biodiesel/kg crude oil is produced using the first option, whereas only 0.85–0.95 kg biodiesel/kg crude oil is obtained using the second option. This difference, about 7%, in the biodiesel yields affects directly the economics of the process. On the other hand, from Figures 3.3 and 3.4, it can be observed that the downstream process to purify the biodiesel is different depending on the type of the selected production process. Therefore, the influence of the hierarchical analysis of the technologies provides a set of criteria to decide the best way to obtain the final product. Thus, the use of a specific technology can change the overall design of a

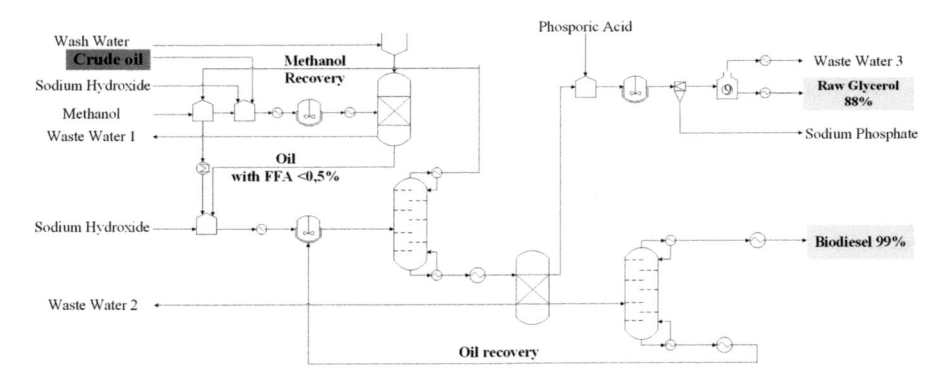

FIGURE 3.3
Biodiesel production using crude oil and alkali-catalyzed process.

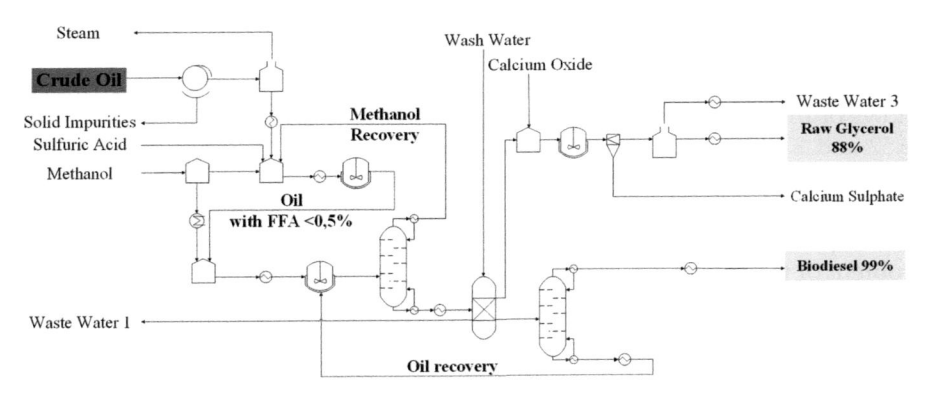

FIGURE 3.4
Biodiesel production using crude oil and acid-catalyzed process.

biorefinery due to changes in the process streams. In this case study, the alkali-catalyzed process also presents a better biodiesel price than the acid-catalyzed process. During this analysis, biodiesel production costs of 0.15 and 0.23 USD/L are obtained from the alkali-catalyzed and acid-catalyzed processes, respectively. Moreover, the fixed capital cost and the operational expenditures are affected. Finally, the impact of the technology selection to obtain the biodiesel from crude oil is also reflected in the environmental analysis. The potential environmental impact (PEI) is better in the alkali-catalyzed process than in the acid-catalyzed process according to the evaluation of the process performed using the WAR algorithm. The abovementioned implications (i.e., yields, economic indicators, and environmental impact) are greater when a biorefinery system is properly designed to achieve the main objectives. Therefore, a careful hierarchical analysis of the technologies must be done to avoid low productivities, high product prices, and high environmental impact, which is the opposite idea of a modern biorefinery.

Another perspective that can be considered after the hierarchical analysis of technologies is the influence of the number of processes included in the biorefinery design. The biorefinery scheme presented by Aristizábal et al. (2015) [10] considers a lignocellulosic feedstock (i.e., coffee cut-stems and sugarcane bagasse) to produce furan-based compounds and energy vectors, such as ethanol, through the simulation of six scenarios. However, to achieve the goal of the hierarchical analysis, only two scenarios will be discussed. The process flow diagram of each scenario is described graphically by Aristizábal et al. (2015) [10], and a brief description is presented in Section 7.4. The base case (scenario 1) considers the production of ethanol and octane. These products were proposed considering the hierarchical decomposition of the feedstocks, which was based on the chemical characterization of the raw material and the potential of the feedstocks to produce high value-added products of the petrochemical industry. Ethanol is produced using glucose extracted from the cellulose fraction, whereas octane is produced using the furfural obtained from the pentose fraction of the raw material. In this scenario, when coffee cut-stems are used as feedstock, the mass flows of the main products were 664.79 and 925.65 kg/h, with sale prices of 0.52 and 2.58 USD/kg for ethanol and octane, respectively. Nevertheless, to understand the influence of increasing the number of processes in a biorefinery scheme, a new product (i.e., nonane) is included. Nevertheless, nonane is produced using a fraction of the glucose obtained from the cellulose content of the raw material. Therefore, it is necessary to specify a restriction in this process to satisfy the degrees of freedom of the overall scheme. Thus, nonane was produced using 40% of the glucose generated in the process (scenario 2). Based on the abovementioned restriction, the mass flows of the main products of the process were 406.01, 925.65, and 232.03 kg/h, with sale prices of 0.68 USD/kg, 2.58 USD/kg, and 10.80 USD/kg for ethanol, octane, and nonane, respectively. In this case study, the inclusion of other products as well as other technologies of conversion (i.e., aldol-condensation, hydrogenation, and dehydrogenation/hydrogenation to liquid alkanes) increases apparently the net income of the overall biorefinery system. Additionally, the amount of equipment and utilities that are necessary to carry out the chemical transformations may have more influence than the revenues generated for the sale of the obtained products. Indeed, the net present value (NPV) of scenario 1 is higher than that of scenario 2. Therefore, in this case study, an increase in the number of products does not improve the economics of the overall biorefinery scheme. In addition, the PEI of scenario 2 is higher than that of scenario 1, which means that the inclusion of nonane as a product decreases the overall sustainability of the biorefinery.

Finally, the hierarchical analysis of technologies as well as the hierarchical decomposition of the feedstocks and products is the first step in the design of a biorefinery system. Therefore, an in-depth study of the chemical characterization of the raw materials, platforms that can be obtained as well as products and technologies can provide an overall idea of the final purpose of the biorefinery. Then, it is possible to understand the main impact from the techno-economic and environmental point of view.

3.2 Sequencing Concept

The term "sequencing" is used to establish a logical order to relate technologies and products. This involves certain restrictions to establish platforms and processing lines (see Figure 3.1B). To do this, it is very important to define the purpose of the biorefinery in terms of products and available technologies. An example can be described in a sugarcane biorefinery that produces sugar (for food applications) and ethanol. The first step is the juice extraction and the second step is to use the extracted juice to obtain other products. At this point, many options will be available, and depending on the biorefinery purpose, the juice can be used to produce sugar or ethanol. It is important to note that ethanol can also be produced from noncrystallized sugar (molasses), which is a by-product of sugar production. If juice is entirely used to produce sugar, ethanol is obtained from molasses, and therefore, the sequence of producing sugar in its first place is evident. In addition, the amount of molasses produced will depend exclusively on the amount of processed raw material (sugarcane) and the efficiency of the technologies used in the production of sugar. Taking this into consideration, many alternatives of processing lines can be defined depending on the targets of the biorefinery even to obtain the same products.

Using the same example of sugar and ethanol from sugarcane, another alternative is to use a 50:50 ratio of juice to produce ethanol and sugar, and molasses from sugar production destined to produce only ethanol. It can be easily inferred that production capacities will affect the economic and environmental performances of a biorefinery at its processing stage. Nevertheless, it should be considered that portfolios with more products will be more sensitive to this type of changes [5]. Additionally, the selection of the most promising processing sequence is a very challenging task in multiproduct portfolios, and in many cases, it is important to use systematic tools to screen technologies and also distribution of platforms (i.e., percentage of one platform going to one process or another) [5]. At this point, different systematic tools such as optimization subroutines can be used for decision-making [6].

Another sequencing principle is that any processing line should include, first, the production of those substances with maximal restrictions in purity to avoid the accumulation of solvents and any contaminants after processing stages. One example is biorefineries designed with the main purpose of obtaining high value-added products (antioxidants, flavors, etc.). In the case of citrus residues, peels are used first for antioxidants, and then the residues undergo pectin production. Finally, the last residues can be used as raw material for ethanol or energy through cogeneration cycles [11]. Likewise, the selling price and the purity restrictions of every product decrease. Usually, any product to be used for food directly or as an additive must appear in the first step of a biorefinery processing line.

3.2.1 Logical Order for Processing Lines

The processing lines in a biorefinery can be considered as the route that the raw materials and their derivatives will follow during the process until they will be converted into the final products. In this sense, different restrictions related with the quality of products, the amount of reactants to be added in the process as well as the quantity of conditioning processes must be considered, aiming to avoid unnecessary steps in the biorefinery.

According to the hierarchical decomposition performed in Section 3.1.1, different types of raw materials can be upgraded in a biorefinery system, which have different potential applications. However, regardless of the feedstock to be employed, the high value-added products that can be obtained through physical processes such as extraction, pressing, or stripping must be performed first aiming to avoid the use of any compound that can make the subsequent steps of the process difficult. For instance, oleaginous materials must be pressed in the first lines of the biorefinery to produce oil with high purity and oil recovery yield. On the other hand, other types of raw materials such as leaves or peels have a large amount of antioxidants to be extracted using an organic solvent [12]. Therefore, before subjecting this kind of raw materials to any process that involves a chemical, thermochemical, or biotechnological transformation, the extraction of this type of compounds must be done. This ensures a maximum extraction yield with low difficult in the separation and purification stage. According to the abovementioned, a logical order must be followed to minimize the amount of processing lines for obtaining the final products of the biorefinery as well as to maximize the overall mass yields of the process.

Once the raw material is stripped, exhausted, or pressed, the subsequent step is to produce the building blocks or platform compounds that are necessary to produce the final products of the biorefinery. This step can include a pretreatment stage to divide the feedstock into its main constituent components (e.g., starch, cellulose, hemicellulose, lignin, protein), if the biorefinery is based on chemical or biotechnological pathways. Nevertheless, the final products are derived from syngas or bio-oil (i.e., thermochemical technologies of transformation) processes such as drying, size reduction, and pelletizing are more suitable. Afterward, the production of bulk or fine products should be performed because their production can involve the use of other chemical agents or special process conditions. As a final point, the energy vector is the last processing line included in the biorefinery system due to its low sale price and the maturity of this type of processes. Therefore, a basic logical order to be considered in the design of a biorefinery could be the following: extraction of value-added products, pretreatment, production of platform compounds, production of chemicals, and generation of energy vectors. This order guarantees minimizing, in some share, the unnecessary use of energy and reactants.

The above-mentioned sequencing description has been applied to different biorefineries reported in the literature [13,14]. One of these cases is the biorefinery reported by Dávila et al. (2017) [15] in which a model biorefinery for avocado (*Persea americana* Mill.) processing is proposed. This raw material has a great potential to be implemented using the biorefinery model to produce a series of products such as oil, protein, phenolic compounds, xylitol, and ethanol. In this case study, it is possible to observe easily the sequencing concept because this process involves all the abovementioned basic logical order of the processing lines. First, the pulp, peel, and seed from avocado are pressed and extracted to produce oil, protein, and phenolic compounds only using a mechanical process for the oil production and a supercritical carbon dioxide technology for the phenolic compound extraction. These processes are carried simultaneously which have the same importance in the biorefinery from the sequencing point of view. Subsequently, based on the exhausted solids from the previous processes, sugars are produced as a platform to be used in the production of the other products. For this reason, a pretreatment stage that involves the implementation of a diluted acid process is carried out. Here, a stream rich in xylose and a solid rich in lignin and cellulose streams were produced. Therefore, the xylose, glucose, and lignin contents of the raw material were used as platforms to produce xylitol, ethanol, and electricity. In this process, these products are manufactured at the same time because the three produced platforms are generated in independent streams at the pretreatment stage. Thus, the chemical production (i.e., xylitol and ethanol) is done using biotechnological processes such as fermentation, whereas the electricity generation is carried out using a cogeneration system where the thermochemical conversion of the lignin fraction through its combustion and heat recovery using a combined Rankine cycle is implemented.

The result of applying the sequencing concept is a "small" process in which all the desired products are obtained. The implementation of this concept also has impact on the economic and environmental indicators. A bad selection of the processing order for different raw materials can make a biorefinery unprofitable due to the increase in the utility consumption, raw material use, and capital investment. Therefore, the sequencing concept can be seen as a conceptual design tool that must be combined with concepts related with process engineering to develop a more detailed route to obtain the proposed products from the first level of the hierarchical decomposition of feedstocks and products.

3.2.2 Relation between Products and Technologies

The processing lines of the biorefinery, which gives an idea of the order for the processes and products, are complemented through the dissertation of the type of technology that will be used to obtain the desired products. Therefore, a direct relation between the products of the biorefinery and the technologies to produce them can be established. Moreover, this relation affects the overall biorefinery from the techno-economic and environmental points of view due to the existence of a lot of technologies and processes that can be used to produce a certain product.

There are different ways to produce the same product in a biorefinery system. For instance, phenolic compounds can be extracted through the use of supercritical fluids, Soxhlet extraction, ultrasound extraction, and organosolv extraction, among others. Even so, the most recommendable technology to be implemented in a biorefinery system is the extraction using low-cost solvents such as ethanol. Other options might have a higher mass yield or an extraction recovery index. However, the economics of the process and its mature in the industrial sector are also an important consideration that must be taken into account. Another example that demonstrates the existing relationship between the products and the technologies used is the lactic acid case. This compound can be produced using different downstream processes. The first one is the crystallization of the lactic acid and the second one is the extraction of lactic acid using organic solvents such as decanol and dodecanol. The first option is more suitable for food application due to the fact that in this process, low quantities of chemicals are used, whereas the second option gives a solution of lactic acid at a needed concentration that can be used to produce polylactic acid. Both technologies are available at the industrial level; thus, the relation between the product and the technology is based on the final use of the product.

Finally, the abovementioned examples can be extrapolated to the biorefinery context. The selection of the technologies and process conditions to obtain the desired products of the biorefinery is also performed if the selected technology is available at the industrial level. Thus, the processes that are implemented in a biorefinery scheme should have some similarities in terms of intermediate products and process conditions aiming to reduce the utilities and equipment costs. This fact is discussed in a better way with other examples in Chapter 7.

3.3 Integration Concept

Integration makes reference to the maximum use of resources within the same plant. As mentioned previously, the different types of biorefineries (first, second, and third generations) can be integrated with each other.

The integration levels are directly related with process sequences and with the hierarchical decomposition of feedstocks, products, and technologies (see Figure 3.1C). Integration emphasizes on the maximum possible use of resources within the same plant. Therefore, it is decomposed into three levels: (i) integration of feedstocks, (ii) integration of technologies, and (iii) integration of products.

3.3.1 Feedstock Integration

Integration of feedstocks is suggested for the use of streams that result from one process and can be directly exploited in other processes (e.g., the exploitation of molasses resulting from the processing of the sugarcane to produce ethanol). In other words, the target is to valorize the streams with lower economic value if they are not transformed into end products [6,16]. Another example of this integration is the use of CO_2-rich streams to cultivate microalgae and obtain metabolites with higher value added [17–19]. Integration of feedstocks can also be seen as an alternative to minimize waste streams. Additionally, it is possible to obtain products to be used as raw materials for further processing in the same biorefinery: ethanol and microalgae oil to produce biodiesel or to obtain antioxidants that can be added later to food products (e.g., sugars or edible oils). At this point, the relationship between sequencing and integration concepts is evident.

The goal must be the integration of first-, second-, and third-generation feedstocks to increase the efficiency of the overall biorefinery. For example, in the case of a sugarcane-based biorefinery, sugarcane juice, molasses, and CO_2 from cogeneration can be fully used as platforms for other products within a common production facility.

Moncada et al. (2016) [20] proposed a biorefinery scheme for the conversion of *Pinus patula* barks in Colombia. The process scheme presented

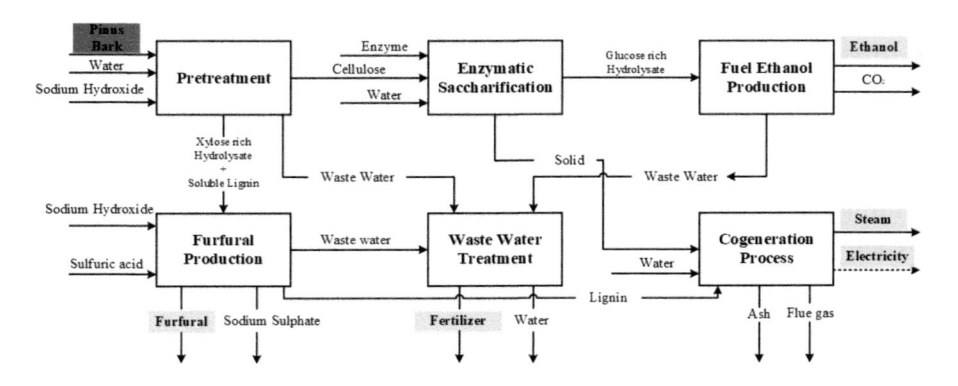

FIGURE 3.5
Simplified block diagram of the biorefinery for the production of fuel ethanol and furfural using *P. patula* bark as a feedstock.

in Figure 3.5 considers the holistic approach adopted for the design of the process for the joint production of ethanol and furfural following the hierarchy, sequencing, and integration concepts proposed previously [20]. The integration involves using streams leaving one block of the biorefinery to another. For example, the sugars obtained in the pretreatment stage (xylose derived from the hemicellulose) are used for furfural production; sugars obtained in the enzymatic saccharification stage (glucose derived from the cellulose) are used as the substrate for ethanol production; the remaining solids and lignin are used in the cogeneration stage to produce steam and electricity. Based on the mass balances presented by the author [20], the yield of the products is 0.10 kg of ethanol per kg of fresh *P. patula* bark fed to the biorefinery; for furfural, this value corresponds to 0.16 kg/kg, and for lignin. 0.71 kg/kg.

3.3.2 Technology Integration

The integration of technologies also plays an important role in the performance of a biorefinery system including three main levels: (i) heat integration, (ii) mass integration, and (iii) process intensification. Heat integration should always be considered because it has economic and environmental impacts of a biorefinery [21,22]. Heat integration can be done using different methodologies, from pinch analysis [1] to contemporary methodologies based on superstructure models [23]. Mass integration is also important because one of the issues in biorefineries is the consumption of water and materials that can be recovered and recycled. Different methodologies used for mass integration can be implemented [24], keeping in mind that mass integration will have economic and environmental impacts [6,21]. There should be a priority standard: integrated technologies over separated technologies. Integrated technologies are better performing technologies than stand-alone technologies. For example, a simultaneous saccharification and fermentation (SSF) is a much better performing technology than stand-alone hydrolysis and fermentation.

3.3.2.1 Energy Integration

This stage aims to take advantage of all the mass and energy streams within the process. Therefore, in this stage, mass and energy balances become transcendental. Mass integration can refer to the optimization of the use of reactants and their recycle. The pervaporation process for separating different components is a good example [25–28]. Energy integration focuses on the design of a heat exchanger network (HEN) or the implementation of power cycles in order to improve the efficiency of each stage in the biorefinery [29].

Heating and cooling are some of the most common operations in biorefineries. A considerable amount of utilities is used in different stages of a biorefinery (reaction, distillation, crystallization, gasification, etc.) to achieve the required operating conditions. To satisfy the need for heating and cooling, biorefineries require the extensive use of energy, and an excessive usage of external heating/cooling utilities can incur in a substantial economic burden and make the process energetically unfeasible (requiring more energy than the one produced) [30]. In this regard, the aim of HENs is to use the energy of the hot and cold streams of a biorefinery/process to supply as much as possible the heating and cooling requirements for the process and provide the remaining heating and cooling with utilities.

Pinch analysis is a methodology to minimize the consumption of energy in processes by calculating and achieving energy targets that are thermodynamically feasible. This methodology was proposed in 1979 by Linnhoff and Flower [31]. This technique is one of the most mature and established methodologies for process integration and improvement. Different publications about this technique have been reported, and some of them are at the industrial level [32–34]. This methodology is considerably simple for application. Through the heat exchange of the hot and cold streams of the system, the requirements of additional utilities decrease considerably. It has been extended to other purposes of integration that are different from energy, such as mass integration [35], water pinch [36–38], and simultaneous integration of mass and HENs [39].

3.3.2.2 Process Intensification

Process intensification based on reaction–reaction (R–R), reaction–separation (R–S), and separation–separation (S–S) is a way to make any plant including biorefineries more compact and effective. The SSF and the extractive distillation have demonstrated the reduction of energy consumption in ethanol production lines [40–47]. Considering the advances that have been made on the genetic modification of microorganisms to add new metabolic routes and, therefore, the ability to assimilate new substrates, a series of modifications have been proposed. The idea is to seek for the reduction of the required stages in fermentation processes and the costs associated with them, as well as better yields in terms of a lower production of inhibitory compounds. The main process configurations are SSF, simultaneous saccharification and co-fermentation (SSCF), and consolidated bioprocessing (CBP) [48].

The SSF is a process that combines the enzymatic hydrolysis with the fermentation to obtain the product in a single step [49]. It has great advantages with respect to other fermentative processes. Some of the advantages are the use of a single equipment for the fermentation and saccharification, reducing both the residence times and the capital costs of the process [50]. Another prominent advantage is the reduction of inhibitory compounds from enzymatic hydrolysis (mainly, sugars inhibiting both the saccharification and consumed *in situ* by the microorganisms), which improves the overall performance of the process [49]. Due to its great advantages, SSF has been widely investigated for the production of ethanol and butanol from lignocellulosic and starchy raw materials [49]. It can have disadvantages that limit its application at the industrial level compared to the SHF [40]. The pH and temperature of the process generally do not agree, and a compromise should be found [40].

The SSCF is based on the use of an enzymatic complex to decompose the cellulose into hexoses at conditions that also solubilize the hemicellulose into pentoses (usually with acids) [51]. These sugars produced *in situ* together with pentoses after pretreatment are consumed simultaneously by a specialized microorganism [52]. SSCF presents great benefits such as the use of less equipment, less reaction time, less pollution, integral use of raw materials, and high efficiencies of production [53]. However, recent advances in SSCF are not enough today to involve this process at the industry level.

The objective must be a single process or stage in which the fermentation product is directly obtained from the cellulose and the production process of the respective enzymes, and the fermentation of the substrate is made by a single microorganism [54]. Therefore, no additional hydrolytic enzyme is required [55]. This process is called consolidated bioprocessing (CBP). With this process scheme, the production cost can be reduced, making the CBP a potential process for fermentative production from lignocellulose [56]. During the past few years, some microorganisms have been evaluated using a CBP scheme. Up to date, *Saccharomyces cerevisiae* is the most advanced option [57]. Some of the difficulties in using *S. cerevisiae* in a CBP are as follows: (i) adverse effects of the co-expression of multiple recombinant proteins, (ii) modulation of simultaneous co-expression of multiple genes at the transcription level, and (iii) misfolding of some of the secretory proteins [58].

Other processes in which reactions and separations are integrated are extractive fermentation or membrane fermentation. Many fermentations are inhibited due to high concentrations of the product. This phenomenon is known as end-product inhibition. This situation has led to look for solutions associated with the *in situ* removal of the product, in order to obtain higher amounts of products. There are two main types of simultaneous fermentation–separation schemes: membrane based and solvent based. Membrane-based technique consists in locating a membrane inside a fermenter that selectively separates the product from the rest of the broth. Pervaporation is one of the most studied examples of this technique [59–62], consisting in coupling a membrane normally composed of polydimethylsiloxane to the fermentation broth, and the membrane selectively separates the product into a vacuum-created vapor phase. In the case of solvent-based schemes, a water-immiscible solvent is added to the fermentation broth in order to create two phases and to remove the product from the culture medium in the organic phase. This scheme implies the separation of the organic phase, recovery of the solvent, and further purification of the products [63]. In some cases, the coupling of the system to a vacuum flask to ease the separation of the volatile compounds is considered. Multiple solvents (*n*-dodecanol, olefin alcohol, polypropylene glycol, and dibutyl phthalate, among others) have been studied [64,65], but the most important feature is that they are 100% biocompatible (nontoxic and no inhibitors). Chapter 4 includes other interesting examples regarding the advantages of using integration in different cases already applied in the industry.

3.3.3 Product Integration

The integration level of products can appear with a final product of a biorefinery section, which can be a raw material in other processing stages. For instance, ethanol that is used for the pectin precipitation can actually be produced from the same solid fraction (lignocellulose) that results in the pectin extraction.

3.3.3.1 Intermediate Products

All the processing lines of the biorefinery include the main or final products as well as some intermediate products that have been processed but require further processing to be ready for the market. The concept of intermediate products is obtained mostly from the definition in chemical reactions, where, for example, some intermediate complexes or molecules are part of some reaction mechanisms before the final product is obtained. Diglycerides and monoglycerides are examples of intermediate products. However, in the current design of biorefineries, building blocks and platform concepts are preferred. Intermediate products cannot be confused with the building blocks or platforms because the latter have a possible market without further processing (e.g., sugars).

3.3.3.2 Building Blocks

In a biorefinery, platforms play an important role as the link between feedstocks and products. The platform products can be transformed to a spectrum of high value-added products through conversion processes that were described in Chapter 1. This concept fulfills the same goal as in the petrochemical industry where the crude oil is fractionated into intermediate products (e.g., naphtha) to generate energy and chemicals. The platform products are also known as pillars, bulk chemicals, bio-based platform molecules, or building block chemicals, and are the base for obtaining value-added products [66–69]. Table 3.1 shows the most important platforms that can be used in the formulation of biorefineries.

TABLE 3.1

Platforms Classification, Description, and Examples

Platform	Description	Example
Syngas	Mix of CO and H_2 obtained by gasification of feedstocks	The base to produce lower alcohols (i.e., methanol), fuel (i.e., Fischer–Tropsch diesel) and chemicals. Syngas can be submitted to a fermentation process to obtain ethanol, methanol, ammonia, etc.
Biogas	Mix of CH_4 and CO_2 obtained by anaerobic digestion of feedstocks	The base to produce hydrogen by steam reforming. The base to produce electricity and heat by combustion.
C6 sugar and C5/ C6 sugars	Glucose, fructose, and galactose obtained by hydrolysis of sucrose, starch, and lignocellulosic biomass. Xylose and arabinose obtained by hydrolysis of hemicellulose.	The base to produce alcohols and organic acids by fermentation. The base to produce biochemicals by reactions of dehydration, hydrogenation, and oxidation.
Plant-based oil and algae oil	Triglycerides generated from oleaginous plants such as palm, coconut, rapeseed, sunflower, soybean, and castor, and microalgae and macroalgae.	The base to produce biodiesel by transesterification. Glycerol as a main co-product of biodiesel is the base to produce propylene glycol, 1,3-propanediol, acrylic acid, and propylene by fermentation and anaerobic digestion.
Organic solutions	Solution rich in carbohydrates, proteins, minerals, enzymes, organic acids, etc., obtained by pressing of wet biomass.	Source of fuels and biochemicals by fermentation. The press cake can be a potential basis to syngas, sugars, and lignin platforms.
Lignin	Component present in the cell walls of plants, and it is estimated that 30% of the organic carbon is present in the biosphere.	Involves obtaining the syngas products (i.e., methanol, ethanol, mixed alcohols, C_1–C_7 gasses), hydrocarbons (i.e., BTX (benzene, toluene, and xylene), cyclohexane styrenes), phenols, oxidized products (i.e., vanillin, aromatic acids, aliphatic acids), and macromolecules (i.e., composites, adhesives, pharmaceuticals).
Pyrolysis oil	Mixture of compounds of different size molecules. Biomass depolymerization at moderate temperatures.	Source of organic acids, phenols, and furfural.

(Continued)

TABLE 3.1 (*Continued*)

Platforms Classification, Description, and Examples

Platform	Description	Example
Furans	These compounds are obtained through chemical transformation of C5 and C6 sugars. Furfural and 5-hydroxymethylfurfural are obtained through the hydrolysis of biomass, followed by a dehydration reaction of glucose or xylose in the presence of an acid catalyst.	The furan compounds are promising molecules as platforms for the generation of chemicals and jet biofuels.
Glycerol	The glycerol is obtained mainly in the biodiesel production process and in bioethanol production in lower proportion.	Glycerol as a substrate can be transformed through catalytic or biological routes into 1,3-propanediol, 1,2-propanediol, succinic acid, propionic acid, acrylic acid, ethanol and co-products, *n*-butanol, etc.

3.4 Sustainability Framework

Considering the interest on sustainable and environmentally friendly processes in the past few decades, there is a necessity of having certain criteria or indicators to determine the performance of a given biorefinery. Many strategies and critical points in sustainability measurement are discussed today. Some basic points are described in Sections 3.4.1 and 3.4.2.

3.4.1 Role of Green Chemistry and Green Engineering in Biorefinery Design

Biorefineries can be extended and integrated within the green chemistry and green engineering principles [70]. A set of guidelines can be proposed in order to integrate the hierarchy, sequencing, and integration concepts discussed earlier. According to the EPA, the green chemistry concept is defined as "the design of chemicals and processes that reduce or eliminate the use or generation of hazardous substances. Green chemistry applies across the life cycle of a chemical product, including its design, manufacture, use, and final disposal." In the same way, green engineering principles are defined as "the design, commercialization, and use of processes and products in a way that minimizes pollution, promotes sustainability, and protects human health without sacrificing economic viability and efficiency." These concepts were well developed by Anastas and Warner in 2000, which are considered as a guide for more sustainable chemical processes [71]. Based on these guidelines, Moncada et al. (2016) extended and categorized the criteria, as described in Sections 3.4.1.1–3.4.1.4 [3].

3.4.1.1 Economics

- Integration of first-, second-, and third-generation feedstocks is the goal. In the case of sugarcane as raw material, sugarcane juice, molasses, and CO_2 from cogeneration can be fully used as platforms for other products within a common production facility. The mass integration favors the biorefinery efficiency.

- Integrated technologies should have priority over separated technologies. For instance, SSF is a much better performing technology than stand-alone fermentation.

- Include as many as possible products in the biorefinery. Stand-alone production usually present in the industry is the worst case. Economic, environmental, and social impacts will always be more positive as the number of products increases [6,7,14,72].

- Innovative engineering solutions. All the state-of-the-art technologies for biorefinery design can be applied. For instance, the combined cycle coupled with gasification technology for cogeneration has a better performance than the technology based on direct combustion. An interesting example is the recovery of lactic acid from fermentation broth using ionic resin exchange columns and reactive distillation [73]. Another case is the production of anthocyanins from cell cultures.

- Reduction of energy consumption and by-products with low added value should be a challenge during the design [74]. Thermodynamic approaches are very effective for this analysis: topological-thermodynamic analysis [75] reduces uncertainties during the design of distillation columns and exergy analysis can give better insights into the cycle efficiency in cogeneration processes [76].

- The use of modern tools and strategies of analysis and evaluation for environmental, technical, and economic impacts is very important. The use of Aspen Plus and other software for specific units and calculations results in an accurate way to predict the behavior of large-scale processes [77,78]. Also, different tools such as the WAR algorithm developed by the EPA [79,80], GREENSCOPE, Aspen Process Economic Analyzer, Aspen Energy Analyzer, and other tools allow having an idea on which biorefining alternative could be a good design in terms of performance from the techno-economic and environmental points of view. In addition, the implementation of optimization strategies and models could be interesting when coupling with further design.

- Design processes for innovative molecules with high added value. Antioxidants, functional sugars, biopolymers, and biosurfactants are good examples for this rule [81].
- Supply chain and logistics should be an essential part of a green biorefinery. After good biorefinery design, the context of the project including transport, product distribution, and others can be involved in the discussion to retrofit the input data if needed.
- Generally, scenarios and sensitivity analysis are included as a main strategy for your design. Examples demonstrated an increasing understanding of your biorefinery possibilities [20].
- The hedging strategies should be considered to avoid unexpected and undesirable changes in prices of raw materials and energy used in the biorefinery. This makes possible to manage financial risks [82]. Then, it is possible to ensure that the NPV and overall production costs are stable and serious results for the biorefinery projects without speculations in the market.
- A simple way is included to understand how efficient your proposed biorefinery is. An example is an index depending on the relationship between the number of products such as the coefficient of the mass to be obtained and the quantity of the biomass used exclusively for this purpose on a wet basis. This index can be called mass index of biorefineries (MI_B), and it is defined by Equation 3.1. The index reflects one principle of sustainability: integral use of the raw materials.

$$MI_B = \frac{\sum_{i=1}^{n} m_i^p}{\sum_{j=1}^{n} m_j^f m} \tag{3.1}$$

where:

    i, j denote the species i and j, referring to the products and feedstocks, respectively.

    m denotes the mass flow rate of feedstocks and products, and superscripts p and f denote the products and feedstocks, respectively.

3.4.1.2 Society

- Assessment of socioeconomic impacts. The supply chain is improved when the biorefinery is working properly in terms of efficiency. This allows to increase incomes and then to redistribute them on the growers or other chain actors. Therefore, the welfare is increased and new jobs can be generated in a sustainable way.
- Food security implications should be analyzed in an objective way, including social, environmental, and economic impacts. In this case,

the biorefinery producing food and functional molecules such as additives should be studied to get the real map of contributions or effects on food security. For instance, xylitol production is projected as a high value-added molecule because it has similar characteristics in taste to sucrose. These types of products can reduce the pressure from society regarding the conflict between food and energy, for example.

3.4.1.3 Environment

- Reduction of waste streams and integration of products with feedstocks in multiprocessing biorefineries for further valorization is a key challenge. For instance, the pinch integration approach for mass and energy in a multiprocessing biorefinery could demonstrate the effectiveness of this rule. Other important example is the integration of cogeneration systems using solid wastes from the same facility. In addition, the energy integration levels are important to reach the maximum energy efficiency levels. Water integration (recovery) prevents the disposal of organic effluents, thus reducing the possible environmental impacts and reducing costs associated with fresh raw material outsourcing.

- Preserve ecosystems and biodiversity. The use of second- and third-generation raw materials helps to improve the overall performance of the biorefinery avoiding the need of other natural sources, which may affect preservation. For instance, the production of fuel alcohol from agroindustrial residues is a challenge, but it will decrease the dependency on first-generation feedstocks. When residues are used, food security questions could be solved to avoid important social implications.

- Adopt the term of life cycle. A biorefinery should look upstream and downstream of its production facility considering sustainable supply chain designs. The efforts for sustainable practices should start from farming and end up in sustainable product consumption and management.

- Design safer processes. The use of chemicals and materials under controllable conditions is a goal of the good simulation to ensure the stability of the units. The technology hierarchy concept plays an important role at this stage because the correct design priority follows the next sections of the process in terms of safety.

3.4.1.4 Green Biorefinery Rules

All people currently working in the area of green chemistry, biorefineries, and sustainable design, among others, are constantly asked to justify the criteria to determine that its design/process/biorefinery is "green." Based on this, Anastas and Warner proposed a set of 12 principles of green chemistry [71], which have become a widely accepted set of criteria for the assessment of the "greenness" of a given process. These criteria are listed as follows:

1. Designers need to strive to ensure that all material and energy inputs and outputs are as inherently nonhazardous as possible.
2. It is better to prevent waste than to treat or clean up waste after if it is formed.
3. Separation and purification operations should be a component of the design framework.
4. System components should be designed to maximize mass, energy, and temporal efficiency.
5. System components should be output pulled rather than input pushed through the use of energy and materials.
6. Embedded entropy and complexity must be viewed as an investment when making design choices on recycle, reuse, or beneficial disposition.
7. Targeted durability, not immortality, should be a design goal.
8. Design for unnecessary capacity or capability should be considered a design flaw. This includes engineering "one-size-fits-all" solutions.
9. Multicomponent products should strive for material unification to promote disassembly and value retention (minimize material diversity).
10. Design of processes and systems must include the integration of interconnectivity with available energy and material flows.
11. Performance metrics include designing for performance in commercial "afterlife."
12. Design should be based on renewable and readily available inputs throughout the life cycle.

Based on these principles, Tang et al. (2008) [70] proposed a summarized version of those principles with the mnemonic IMPROVEMENTS— PRODUCTIVELY, which covers most of the key issues of green and sustainable chemistry and processing. These mnemonics are shown in Figure 3.6.

At this point, it is very important to take into account that biorefinery rules include the important aspects of the sustainability pillars (environment, society, and economy). In this way, the proposed rules could be considered in the sustainability framework. The biorefineries that follow these principles are considered as sustainable biorefineries or green biorefineries as shown in Figure 3.7.

Principles of Green Chemistry	
Prevent wastes	Inherently non-hazardous and safe
Renewable materials	Minimize material diversity
Omit derivation steps	Prevention instead of treatment
Degradable chemical products	Renewable material and energy inputs
Use safe synthetic methods	Output-led design
Catalytic reagents	Very simple
Temperature, Pressure ambient	Efficient use of mass, energy, space & time
In-Process Monitoring	Meet the need
Very few auxiliary substances	Easy to separate by design
E-factor, maximize feed in product	Networks for exchange of local mass & energy
Low toxicity of chemical products	Test the life cycle of the design
Yes it's safe	Sustainability throughout product life cycle

FIGURE 3.6
Summarized version of the 12 principles of green chemistry. (Taken from [70].)

3.4.2 Design Steps: Role of Process Simulation, Experimental Assessment, and Pilot Testing on Biorefinery Design

The strategy for designing biorefineries involves modeling and simulation based on theories and literature, experimental assessment for verifying yields or providing data as well as laboratory and pilot work to adjust the models and future scale-up.

Laboratory-scale experiments have two main purposes. The first is to validate the production yields reported in the literature for these products. It is important to highlight that the purpose of this stage is to evaluate the optimal conditions to obtain the products or validate certain conditions reported in the literature. The second is to learn about the handling of products and reactants, operation conditions, specific requirements, state of the products, and reactants in order to achieve a better understanding of the process. This "learning" of the process decreases the uncertainties. Then, the results obtained in the experimental stage (yields and operation condition, among others) and the characterization of raw materials and products can be used to simulate the processes and design a biorefinery.

The experimental stage can be done simultaneously with the process simulation. The simulation of a process is a very useful tool to analyze the performance of the process, allowing the modification of variables and conditions that might not be modified or considered at laboratory scale. In addition, the versatility and adaptability of simulated processes can save considerable efforts, time, and reactants, when modifying certain design variables and analyzing the performance of the process. In addition, simulation allows performing scenario and sensitivity analysis as a projective tool on

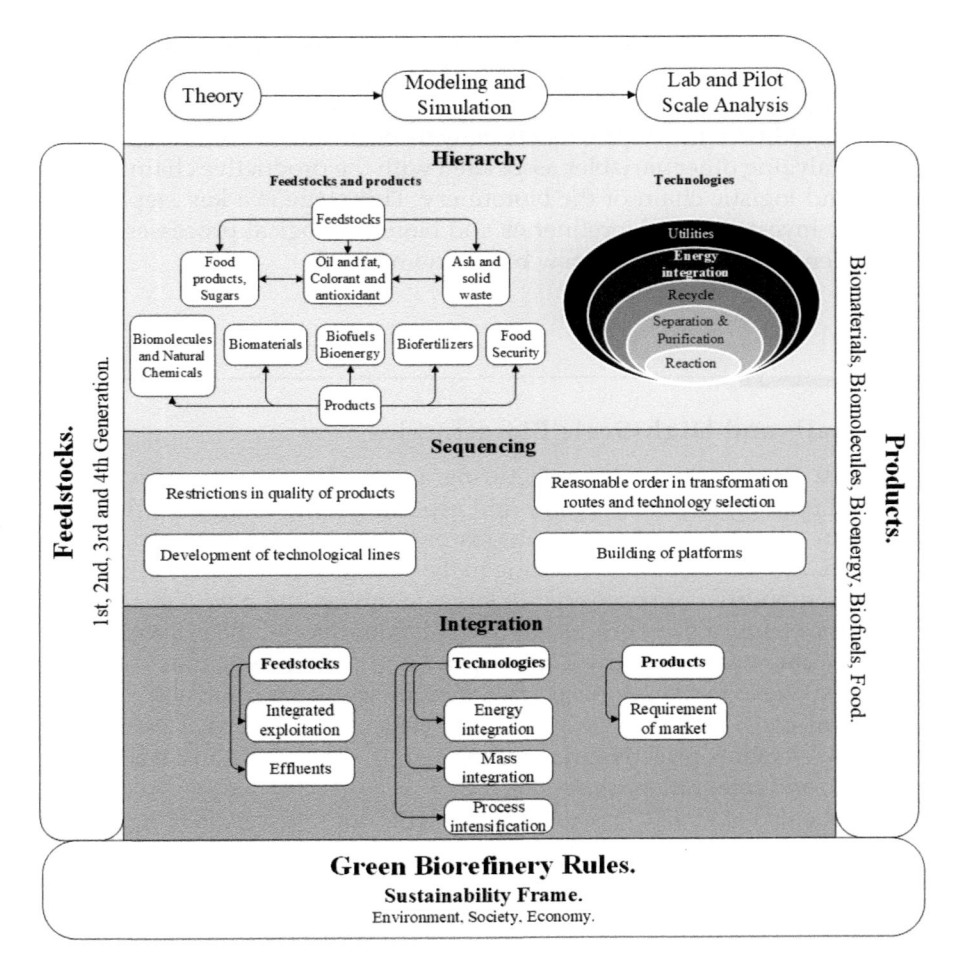

FIGURE 3.7
Green biorefinery structure. Relation of products, feedstocks, concepts, knowledge, and green rules. (Taken from [4].)

the variables, processing scales, and so on. This information will be further explained and detailed in Chapter 4.

Pilot testing is the next stage that can be performed. This stage aims to verify the models, operating conditions, and behavior of the process in a scale considerable higher than the laboratory scale. As the scale of the experiments increases, there is a change in the behavior of the mixing regimes, heat and mass transfer, dimensions and arrangements of the equipment, and requirement of utilities; therefore, it is necessary to reanalyze the performance of the process with this new scale. The scope of this stage is to adapt the process to industrial scales. Demonstration plants are considered to be the final stage before reaching a full industrial process. The objective of these plants is to raise awareness to understand the potential of a given raw material/process, as well as to attract investors to an almost fully operational biorefinery. In addition, the requirements of feedstocks and utilities for this stage allow analyzing other variables associated with the productive chain, supply chain, and logistic chain of the biorefinery. This stage is a key step to promote the investment in biorefineries and biotechnological processes, which strengthen the application of new bio-economies.

3.5 Small- and High-Scale Biorefineries

Generally, the process of manufacturing a given product depends on the required quantity. In the industry and especially in biorefineries, there are two main operation modes: continuous or batch. In addition, products are classified into three groups according to the production volume. Bulk chemicals or commodities are produced in large quantities and have a relative low price. Fine chemicals are produced in small amounts and the price is higher. Specialty chemicals have the highest price and their production volume is the lowest due to their specific use [83]. Biorefinery size is influenced by environmental considerations, governmental policies, and market conditions, and these elements directly influence the supply chain decisions, technology selection, and integration levels [84].

Serna-Loaiza et al. (2017) present some definitions of scale [83]. According to the supply of raw material, the definition of large and small scales is strongly associated with the availability of a given raw material. In this sense, the scale could be assimilated to the *processing scale*, which refers to the *amount of raw material processed in a period*. According to the flow of products, the definition of large and small scales is strongly associated with the type of product that is obtained. Large scale in this case refers to high amounts of product that can be obtained in the process per unit of time. In this sense, the scale could be assimilated to the *production scale*, which refers to the *amount of product obtained in a period*. A third definition of scale can be associated with an economic approach, and it aims at determining the *minimum processing scale at which the process presents a feasible economic performance* [83,85]. It is necessary to perform an economic evaluation of the process and a sensitivity analysis of how it changes the economic performance of the process (NPV) with the flow of raw material and identify the equilibrium point (NPV=0), this would be the start point to determine a "feasible small scale."

Therefore, a biorefinery that processes raw materials such as sugarcane, palm, and corn, and their respective wastes could be considered as large scale, because the amount of feedstocks available is very high. A biorefinery that produces bulk chemicals such as ethanol, sugar, butanol, and biodiesel could be considered as large scale. On the other hand, a biorefinery that produces lactic acid, PHB, and citric acid could be considered as small- to medium-scale processing. They depend on the amount of feedstock; the obtained product and products nature (specialty/fine chemicals). This is why feedstocks such as sugarcane and palm are used in large-scale biorefineries to obtain bulk chemicals such as biodiesel, ethanol, and electricity, because there is a high amount of feedstock available for their use. However, fruit peels and feedstocks in lower amounts are generally used for the production of specialty chemicals that have higher benefit.

Regarding process performance, it is possible to observe two remarked differences associated with the scale. In both small and large scales, generally the most representative costs are utilities and raw materials. This is the case of multiple biorefineries designed with raw materials (castor bean, microalgae, residual banana, olive stone, *P. patula*, and glycerol, among others) for the production of a wide variety of products (energy, ethanol, butanol, PHB, biodiesel, and antioxidants, among others) [10,14,20,46,86–89]. However, the economic share that represents these two costs is not the same for both scales. For example, in small-scale biorefineries, the cost of both items tends to be equal, whereas in large-scale biorefineries, the cost of raw materials tends to be much higher than utilities. Economies of scale can be used to explain this behavior given that as scale increases, the cost per unit of output decreases because fixed costs are divided into more units of output [2]. For small-scale biorefineries, the cost of raw materials usually equals the cost of utilities. This can be explained because utilities must be fully supplied and the integration networks have not significant incidence in their supply. For large-scale biorefineries, feedstock requirements increase and so does the cost, but the exchange networks significantly decrease the requirements of utilities and their respective cost.

3.6 Conclusions

The strategy to design sustainable biorefineries is based on a heuristic approach applying the hierarchy, sequencing, and integration concepts aiming to establish the better way to use different types of feedstocks and to obtain high value-added products. Moreover, the implementation of different technologies and integration between processing lines is the way to increase the productivity and feasibility of a biorefinery scheme. Therefore, the approach discussed in this chapter gives a guideline with examples about how to propose a biorefinery scheme. On the other hand, the biorefinery sustainability is related directly with techno-economic and environmental indicators as well as its potential to impact the communities at all levels.

References

1. R. Smith, *Chemical Process Design and Integration*, McGraw-Hill. 2005.
2. J. M. Douglas, *Conceptual Design of Chemical Processes*, New York, NY: McGraw-Hill. 1988.
3. J. Moncada, V. Aristizábal, and C. A. Cardona, "Design strategies for sustainable biorefineries," *Biochem. Eng. J.*, vol. 116, pp. 122–134, 2016.
4. J. Moncada Botero, "Design and evaluation of sustainable biorefineries from feedstock in tropical regions," Master Thesis. Universidad Nacional de Colombia. Departamento de Ingeniería Química, 2012.
5. J. Moncada, L. G. Matallana, and C. A. Cardona, "Selection of process pathways for biorefinery design using optimization tools: A Colombian case for conversion of sugarcane bagasse to ethanol, poly-3-hydroxybutyrate (PHB), and energy," *Ind. Eng. Chem. Res.*, vol. 52, no. 11, pp. 4132–4145, 2013.
6. J. Moncada, J. A. Tamayo, and C. A. Cardona, "Integrating first, second, and third generation biorefineries: Incorporating microalgae into the sugarcane biorefinery," *Chem. Eng. Sci.*, vol. 118, pp. 126–140, 2014.
7. J. Moncada, J. Tamayo, and C. A. Cardona, "Evolution from biofuels to integrated biorefineries: Techno-economic and environmental assessment of oil palm in Colombia," *J. Clean. Prod.*, vol. 81, pp. 51–59, 2014.
8. J. A. Dávila, M. Rosenberg, and C. A. Cardona, "A biorefinery approach for the production of xylitol, ethanol and polyhydroxybutyrate from brewer's spent grain," *AIMS Agric. Food*, vol. 1, no. 1, pp. 52–66, 2016.
9. L. E. Rincón, J. J. Jaramillo, and C. A. Cardona, "Comparison of feedstocks and technologies for biodiesel production: An environmental and techno-economic evaluation," *Renew. Energy*, vol. 69, pp. 479–487, 2014.
10. V. Aristizábal, Á. Gómez, and C. A. Cardona, "Biorefineries based on coffee cut-stems and sugarcane bagasse: Furan-based compounds and alkanes as interesting products," *Bioresour. Technol.*, vol. 196, pp. 480–489, 2015.
11. J. Moncada Botero, "Design and evaluation of sustainable biorefineries from feedstocks in tropical regions," Master Thesis. Departamento de Ingeniería Química, Universidad Nacional de Colombia sede Manizales, 2012.
12. J. A. Dávila, M. Rosenberg, and C. A. Cardona, "A biorefinery for efficient processing and utilization of spent pulp of Colombian Andes Berry (*Rubus glaucus* benth): Experimental, techno-economic and environmental assessment," *Bioresour. Technol.*, vol. 223, pp. 227–236, 2016.
13. C. A. Cardona Alzate, J. C. Solarte-Toro, and Á. G. Peña, "Fermentation, thermochemical and catalytic processes in the transformation of biomass through efficient biorefineries," *Catal. Today*, vol. 302, pp. 61–72, 2018.
14. J. Moncada, C. A. Cardona, and L. E. Rincón, "Design and analysis of a second and third generation biorefinery: The case of castorbean and microalgae," *Bioresour. Technol.*, vol. 198, pp. 836–43, Dec. 2015.
15. J. A. Dávila, M. Rosenberg, E. Castro, and C. A. Cardona, "A model biorefinery for avocado (*Persea americana* mill.) processing," *Bioresour. Technol.*, vol. 243, pp. 17–29, 2017.
16. A. K. Neu, D. Pleissner, K. Mehlmann, R. Schneider, G. I. Puerta-Quintero, and J. Venus, "Fermentative utilization of coffee mucilage using *Bacillus coagulans* and investigation of down-stream processing of fermentation broth for optically pure l(+)-lactic acid production," *Bioresour. Technol.*, vol. 211, pp. 398–405, 2016.

17. J. J. Jaramillo, J. M. Naranjo, and C. A. Cardona, "Growth and oil extraction from *Chlorella vulgaris* : A techno- economic and environmental assessment," *Ind. Eng. Chem. Res.*, vol. 51, pp. 10503–10508, 2012.

18. F. G. Acién, J. M. Fernández, J. J. Magán, and E. Molina, "Production cost of a real microalgae production plant and strategies to reduce it," *Biotechnol. Adv.*, vol. 30, no. 6, pp. 1344–1353, 2012.

19. F. Gabriel Acien Fernandez, C. V. González-López, J. M. Fernández Sevilla, and E. Molina Grima, "Conversion of CO_2 into biomass by microalgae: How realistic a contribution may it be to significant CO_2 removal?" *Appl. Microbiol. Biotechnol.*, vol. 96, no. 3, pp. 577–586, 2012.

20. J. Moncada, C. A. Cardona, J. C. Higuita, J. J. Vélez, and F. E. López-Suarez, "Wood residue (*Pinus patula* bark) as an alternative feedstock for producing ethanol and furfural in Colombia: experimental, techno-economic and environmental assessments," *Chem. Eng. Sci.*, vol. 140, pp. 309–318, Feb. 2016.

21. S. I. Mussatto, J. Moncada, I. C. Roberto, and C. A. Cardona, "Techno-economic analysis for brewer's spent grains use on a biorefinery concept: The Brazilian case," *Bioresour. Technol.*, vol. 148, pp. 302–310, 2013.

22. J. Han, J. S. Luterbacher, D. M. Alonso, J. a Dumesic, and C. T. Maravelias, "A lignocellulosic ethanol strategy via nonenzymatic sugar production: Process synthesis and analysis," *Bioresour. Technol.*, vol. 182, pp. 258–66, Apr. 2015.

23. K. F. Huang and I. A. Karimi, "Efficient algorithm for simultaneous synthesis of heat exchanger networks," *Chem. Eng. Sci.*, vol. 105, pp. 53–68, Feb. 2014.

24. M. M. El-Halwagi, "Chapter 14: Overview of optimization," in *Sustainable Design through Process Integration. Fundamentals and Applications to Industrial Pollution Prevention, Resource Conservation, and Profitability Enhancement*, pp. 255–286, Amsterdam, Netherlands: Elsevier, 2012.

25. S. Roffler, H. Blanch, and C. Wilke, "Extractive fermentation of acetone and butanol: Process design and economic evaluation," *Biotechnol. Prog.*, vol. 3, pp. 131–140, 1987.

26. N. Qureshi, D. Hodge, and A. Vertes, *Biorefineries: Integrated Biochemical Processes for Liquid Biofuels*, 1st ed., Burlington, MA: Elsevier, 2014, pp. 1–296.

27. G. Agrimi, I. Pisano, M. A. Ricci, and L. Palmieri, "Microbial strain selection and development for the production of second generation of bioethanol," in *Biorefineries: An Introduction*, M. Aresta, A. Dibenedetto, and F. Dumeignil, Eds., Berlin: De Gruyter, 2015.

28. Y. Wang, N. Widjojo, P. Sukitpaneenit, and T.-S. Chung, "Membrane pervaporation," in *Separation and Purification Technologies in Biorefineries*, S. Ramaswamy, H. J. Huang, and B. V. Ramarao, Eds., pp. 259–301, Chichester, UK: Wiley, 2013.

29. K. J. Ptasinski, *Efficiency of Biomass Energy: An Exergy Approach to Biofuels, Power, and Biorefineries*, Hoboken, NJ: Wiley, 2016, pp. 1–756.

30. M. M. El-Halwagi, "Chapter 7: Heat Integration," in *Sustainable Design through Process Integration*, pp. 147–163, Amsterdam: Elsevier, 2012, pp. 147–163, United States.

31. B. Linnhoff and J. R. Flower, "Synthesis of heat exchanger networks. Part II: Evolutionary generation of networks with various criteria of optimality," *AIChE J.*, vol. 24, no. 4, pp. 642–654, 1978.

32. I. C. Kemp, *Pinch Analysis and Process Integration—A User Guide on Process Integration for the Efficient Use of Energy*, 2nd ed. Burlington, MA: IChemE, 2007.

33. B. Linnhoff and D. R. Vredeveld, "Pinch technology has come of age," *Chem. Eng. Prog.*, vol. July, pp. 33–40, 1984.

34. H. Korner, "Optimal use of energy in the chemical industry," *Chem. Ing. Tech.*, vol. 60, no. 7, pp. 511–518, 1988.
35. M. M. El-Halwagi and V. Manousiouthakis, "Synthesis of mass exchange networks," *AIChE J.*, vol. 35, no. 8, pp. 1233–1244, 1989.
36. Y. P. Wang and R. Smith, "Wastewater minimisation," *Chem. Eng. Sci.*, vol. 49, no. 7, pp. 981–1006, 1994.
37. N. Hallale, "A new graphical targeting method for water minimisation," *Adv. Environ. Res.*, vol. 6, no. 3, pp. 377–390, 2002.
38. R. Prakash and U. V. Shenoy, "Targeting and design of water networks for fixed flowrate and fixed contaminant load operations," *Chem. Eng. Sci.*, vol. 60, no. 1, pp. 255–268, 2005.
39. L. Liu, J. Du, M. M. El-Halwagi, J. M. Ponce-Ortega, and P. Yao, "A systematic approach for synthesizing combined mass and heat exchange networks," *Comput. Chem. Eng.*, vol. 53, pp. 1–13, 2013.
40. K. Olofsson, M. Bertilsson, and G. Lidén, "A short review on SSF—An interesting process option for ethanol production from lignocellulosic feedstocks," *Biotechnol. Biofuels*, vol. 1, no. 7, pp. 1–14, 2008.
41. R. Morales-Rodriguez, K. V. Gernaey, A. S. Meyer, and G. Sin, "A mathematical model for simultaneous saccharification and co-fermentation (SSCF) of C6 and C5 sugars," *Chinese J. Chem. Eng.*, vol. 19, pp. 185–191, Apr. 2011.
42. W. R. Gibbons and S. R. Hughes, "Integrated biorefineries with engineered microbes and high-value co-products for profitable biofuels production," *Vitr. Cell. Dev. Biol.—Plant*, vol. 45, no. 3, pp. 218–228, Apr. 2009.
43. O. J. Sánchez and C. A. Cardona, "Trends in biotechnological production of fuel ethanol from different feedstocks," *Bioresour. Technol.*, vol. 99, pp. 5270–95, Sep. 2008.
44. C. A. Cardona, O. J. Sanchez, and L. F. Gutierrez, *Process Synthesis for Fuel Ethanol Production*. Boca Raton, FL: CRC Press, 2009, pp. 1–390.
45. J. A. Quintero, J. Moncada, and C. A. Cardona, "Techno-economic analysis of bioethanol production from lignocellulosic residues in Colombia: A process simulation approach," *Bioresour. Technol.*, vol. 139, pp. 300–307, Jul. 2013.
46. L. F. Gutiérrez, O. J. Sánchez, and C. A. Cardona, "Process integration possibilities for biodiesel production from palm oil using ethanol obtained from lignocellulosic residues of oil palm industry," *Bioresour. Technol.*, vol. 100, no. 3, pp. 1227–1237, Feb. 2009.
47. C. A. Cardona, C. A. García, and S. Serna-Loaiza, "Bioethanol production: Advances in technologies and raw materials," in *Bioenergy and Biofuels*, O. Konur, Ed., p. 523. Boca Raton, FL: CRC Press, 2018.
48. M. Ask et al., "Challenges in enzymatic hydrolysis and fermentation of pretreated *Arundo donax* revealed by a comparison between SHF and SSF," *Process Biochem.*, vol. 47, no. 10, pp. 1452–1459, 2012.
49. M. Ballesteros, J. M. Oliva, M. J. Negro, P. Manzanares, and I. Ballesteros, "Ethanol from lignocellulosic materials by a simultaneous saccharification and fermentation process (SFS) with *Kluyveromyces marxianus* CECT 10875," *Process Biochem.*, vol. 39, no. 12, pp. 1843–1848, 2004.
50. J. M. Oliva, M. Ballesteros, and L. Olsson, "Comparison of SHF and SSF processes from steam-exploded wheat straw for ethanol production by xylose-fermenting and robust glucose-fermenting *Saccharomyces cerevisiae* strains," *Biotechnol. Bioeng.*, vol. 100, no. 6, pp. 1122–1131, 2008.

51. J. Zhang, X. Shao, O. V. Townsend, and L. R. Lynd, "Simultaneous saccharification and co-fermentation of paper sludge to ethanol by *Saccharomyces cerevisiae* RWB222—Part I : Kinetic modeling and parameters," *Biotechnol. Bioeng.*, vol. 104, no. 5, pp. 920–931, 2009.

52. K. Olofsson, M. Wiman, and G. Lidén, "Controlled feeding of cellulases improves conversion of xylose in simultaneous saccharification and co-fermentation for bioethanol production," *J. Biotechnol.*, vol. 145, pp. 168–175, 2010.

53. M. Jin, M. W. Lau, V. Balan, and B. E. Dale, "Two-step SSCF to convert AFEX-treated switchgrass to ethanol using commercial enzymes and *Saccharomyces cerevisiae* 424A (LNH-ST)," *Bioresour. Technol.*, vol. 101, no. 21, pp. 8171–8178, 2010.

54. W. H. van Zyl, "Consolidated bioprocessing for bioethanol production using *Saccharomyces cerevisiae*," *Adv Biochem Eng Biotechnol.*, vol. 108, pp. 205–235, 2007.

55. N. Khramtsov et al., "Industrial yeast strain engineered to ferment ethanol from lignocellulosic biomass," *Bioresour. Technol.*, vol. 102, no. 17, pp. 8310–8313, 2011.

56. L. R. Lynd, P. J. Weimer, G. Wolfaardt, and Y.-H. P. Zhang, "Cellulose hydrolysis by *Clostridium thermocellum*: A microbial perspective," *Cellulosome*, V. Uversky and Irina A. Kataeva, Eds., New York: Nova Science Publishers, 2006.

57. D. G. Olson, J. E. McBride, A. J. Shaw, and L. R. Lynd, "Recent progress in consolidated bioprocessing," *Curr. Opin. Biotechnol.*, vol. 23, no. 3, pp. 396–405, 2012.

58. Q. Xu, A. Singh, and M. E. Himmel, "Perspectives and new directions for the production of bioethanol using consolidated bioprocessing of lignocellulose," *Curr. Opin. Biotechnol.*, vol. 20, no. 3, pp. 364–371, 2009.

59. S. Fan et al., "Enhanced ethanol fermentation in a pervaporation membrane bioreactor with the convenient permeate vapor recovery," *Bioresour. Technol.*, vol. 155, pp. 229–234, 2014.

60. C. Fu et al., "Ethanol fermentation integrated with PDMS composite membrane: An effective process," *Bioresour. Technol.*, vol. 200, pp. 648–657, 2016.

61. S. Fan et al., "Inhibition effect of secondary metabolites accumulated in a pervaporation membrane bioreactor on ethanol fermentation of *Saccharomyces cerevisiae*," *Bioresour. Technol.*, vol. 162, pp. 8–13, 2014.

62. B. Van Der Bruggen and P. Luis, "Pervaporation as a tool in chemical engineering: A new era?" *Curr. Opin. Chem. Eng.*, vol. 4, pp. 47–53, 2014.

63. R. Palacios-Bereche, A. Ensinas, M. Modesto, and S. A. Nebra, "New alternatives for the fermentation process in the ethanol production from sugarcane: Extractive and low temperature fermentation," *Energy*, vol. 70, pp. 595–604, 2014.

64. M. Minier and G. Goma, "Ethanol production by extractive fermentation," *Biotechnol. Bioeng.*, vol. 24, no. 7, pp. 1565–1579, 1982.

65. F. Kollerup and A. Daugulis, "Ethanol production by extractive fermentation–solvent identification and prototype development," *Can. J. Chem.*, vol. 64, pp. 598–606, August, 1986.

66. F. Cherubini, "The biorefinery concept: Using biomass instead of oil for producing energy and chemicals," *Energy Convers. Manag.*, vol. 51, no. 7, pp. 1412–1421, 2010.

67. F. Cherubini et al., "Toward a common classification approach for biorefinery systems," *Biofuels, Bioprod. Biorefining*, vol. 3, pp. 1–13, 2009.

68. IEA Bioenergy Task42, "Biobased chemicals—Value added products from bio-refineries," 2011.

69. V. Aristizábal Marulanda, C. D. Botero, and C. A. Cardona, "Chapter 4: Thermochemical, biological, biochemical, and hybrid conversion methods of bio-derived molecules into renewable fuels," in *Advanced Bioprocessing for Alternative Fuels, Biobased Chemicals, and Bioproducts*, p. In press, Elsevier, 2017.

70. S. Y. Tang, R. A. Bourne, R. L. Smith, and M. Poliakoff, "The 24 principles of green engineering and green chemistry: 'Improvements productively,'" *Green Chem.*, vol. 10, no. 3, p. 268, 2008.

71. P. T. Anastas and J. C. Warner, *Green Chemistry: Theory and Practice*. New York: Oxford University Press, 1998.

72. L. E. Rincón, J. Moncada, and C. A. Cardona, "Analysis of potential technological schemes for the development of oil palm industry in Colombia: A biorefinery point of view," *Ind. Crops Prod.*, vol. 52, pp. 457–465, Jan. 2014.

73. R. Taylor and R. Krishna, "Modelling reactive distillation," *Chem. Eng. Sci.*, vol. 55, no. 22, pp. 5183–5229, Nov. 2000.

74. E. Durand, J. Lecomte, and P. Villeneuve, "From green chemistry to nature: The versatile role of low transition temperature mixtures," *Biochimie*, vol. 120, pp. 119–123, 2016.

75. J. A. Posada Duque, "Design and analysis of technological schemes for glycerol conversion to added value products," PhD Thesis. Departamento de Ingeniería Eléctrica, Electrónica y Computación. Univerisidad Nacional de Colombia sede Manizales, 2011.

76. H. Kencse and P. Mizsey, "Methodology for the design and evaluation of distillation systems: Exergy analysis, economic features and GHG emissions," *AIChE J.*, vol. 56, no. 7, pp. 1776–1786, 2010.

77. E. Gnansounou, P. Vaskan, and E. R. Pachón, "Comparative techno-economic assessment and LCA of selected integrated sugarcane-based biorefineries," *Bioresour. Technol.*, vol. 196, pp. 364–375, 2015.

78. H. J. Huang, S. Ramaswamy, W. W. Al-Dajani, and U. Tschirner, "Process modeling and analysis of pulp mill-based integrated biorefinery with hemicellulose pre-extraction for ethanol production: a comparative study," *Bioresour. Technol.*, vol. 101, no. 2, pp. 624–631, Jan. 2010.

79. D. Young, R. Scharp, and H. Cabezas, "The waste reduction (WAR) algorithm: environmental impacts, energy consumption, and engineering economics," *Waste Manag.*, vol. 20, no. 8, pp. 605–615, Dec. 2000.

80. H. Cabezas, J. C. Bare, and S. K. Mallick, "Pollution prevention with chemical process simulators: The generalized waste reduction (WAR) algorithm—Full version," *Comput. Chem. Eng.*, vol. 23, no. 4–5, pp. 623–634, 1999.

81. M. Henkel *et al.*, "Rhamnolipids as biosurfactants from renewable resources: Concepts for next-generation rhamnolipid production," *Process Biochem.*, vol. 47, no. 8, pp. 1207–1219, 2012.

82. S. Paul and E. H. Mahmoud, *Integrated Biorefineries: Design Analysis and Optimization* (Green Chemistry and Chemical Engineering). Boca Raton, FL: CRC Press, 2012.

83. S. Serna Loaiza, G. Aroca, and C. A. Cardona, "Small-scale biorefineries: Future and perspectives," in *Biorefineries: Concepts, Advancements and Research*, I. Torres, Ed., pp. 39–72. New York: Nova Science Publishers, 2017.

84. A. Kelloway and P. Daoutidis, "Process synthesis of biorefineries: Optimization of biomass conversion to fuels and chemicals," *Ind. Eng. Chem. Res.*, vol. 53, pp. 5261–5273, 2014.

85. C. A. Cardona, "Evaluación de la producción de diferentes tipos de bioenergía para uso a nivel industrial y rural," in *II Reunión Nacional de la Red Temática en Bioenergía—XI Reunión Nacional de la REMBIO*, 2015.

86. J. M. Naranjo, C. A. Cardona, and J. C. Higuita, "Use of residual banana for polyhydroxybutyrate (PHB) production: Case of study in an integrated biorefinery," *Waste Manag.*, vol. 34, pp. 2634–2640, 2014.

87. V. Hernández, J. M. Romero-García, J. A. Dávila, E. Castro, and C. A. Cardona, "Techno-economic and environmental assessment of an olive stone based biorefinery," *Resour. Conserv. Recycl.*, vol. 92, pp. 145–150, 2014.

88. A. A. González, J. Moncada, A. Idarraga, M. Rosenberg, and C. A. Cardona, "Potential of the amazonian exotic fruit for biorefineries: The *Theobroma bicolor* (Makambo) case," *Ind. Crops Prod.*, vol. 86, pp. 58–67, 2016.

89. J. A. Posada, L. E. Rincón, and C. A. Cardona, "Design and analysis of biorefineries based on raw glycerol: addressing the glycerol problem," *Bioresour. Technol.*, vol. 111, pp. 282–293, May 2012.

4

Techno-economic Analysis of Biorefineries

This chapter aims to present the methodology used for the design of biorefinery networks based on the previous concepts of hierarchy, sequencing, and integration approaches. A detailed explanation of the process simulation, mass, and energy integration along with the techno-economic assessment of biorefineries is presented. Different biorefinery concepts are shown in order to present an example of the proposed methodology.

4.1 Modeling and Simulation

The simulation of biorefineries is performed aiming to specify and evaluate the mass and energy balances of the process configurations using different simulation tools. There is a wide variety of process simulators that are intended for the simulation of chemical processes such as CHEMCAD (Chemstations, Houston, Texas, USA), SuperPro Designer (Intelligen Inc., Scotch Plains, New Jersey, USA), ASSETT (Kongsberg Oil & Gas Technologies AS, Asker, Norway), PRO/II (SimSci, Rueil-Malmaison, France), and Aspen Plus (Aspen Technology Inc., Bedford, Massachusetts, USA), among others. Nevertheless, the Aspen Plus software is the most widely used software for the evaluation of biorefinery approaches involving the use of biomass [1–4]. The purpose of the Aspen Plus software is to model and predict the performance of a process, which involves the decomposition of the process into its constituent elements (e.g., units) for an individual study of performance. Additionally, this computer-aided software uses the underlying physical relationships (e.g., material and energy balances, thermodynamic equilibrium, rate equations) to predict process performance (e.g., stream properties, operating conditions, and equipment sizes). The interface of the software can be divided into inputs (i.e., components [chemicals], compounds properties, thermodynamic models, flow rates of raw materials and/or reagents, and operative conditions of the simulated equipment, among others) and outputs (i.e., products and/or wastes flow rates). The data required for the simulation of the different technological schemes, especially the physical–chemical properties of the involved substances, and the design and operation parameters of each unit operation, can be obtained from secondary sources (articles, technical reports, data bases, patents, and others) or from experimental work. One of the most important issues to be considered during the simulation is the appropriate selection of the thermodynamic models that describe the liquid and vapor phases. There are reports available in open literature, which explain the criteria that must be considered for the selection of the proper thermodynamic model that fits better the simulated process. Different criteria such as the type of process (chemical, electrochemical, hydrocarbons, salts, etc.), pressure, and temperature are commonly used in the selection of the thermodynamic model. The Aspen Plus software has incorporated an assistant tool for the selection of the method that allows a good choice of proper thermodynamic models based on the previous criteria for a specific process.

The detailed level of the flow sheets can be improved through the rigorous simulation of the most relevant unit operations (pretreatment, reaction, and separation stages). This requires the utilization of kinetic models for some of the transformation processes, particularly for biological (i.e., fermentation) and chemical transformations (i.e., steam explosion, organosolv, ammonia fiber explosion (AFEX), acid and alkaline hydrolysis). The Aspen Plus reaction modules do not contemplate the introduction of specific kinetic relationships related to the fermentation and enzymatic processes, and thus, different approaches must be done in order to consider the particularities of the biological transformation. In this sense, different computational tools can be used to model the profiles of the different pretreatment and fermentation processes such as Aspen Plus-Excel and Aspen Plus–MATLAB® interfaces. The kinetic models can be directly or indirectly integrated in the simulator depending on the computational tool used. The selection of the proper model to be employed during the subsequent steps of the synthesis procedure is carried out based on the abovementioned information and experimental data reported in the literature. In some cases, accessible information from existing industrial processes was used.

Additionally, regarding process simulators, different methodologies to generate and identify optimal biorefinery networks have been developed. The optimization of superstructures consists of tools and methods including databases, models, superstructure, and solution strategies to represent, describe, and evaluate processing network alternatives. Normally, the optimization of the network is formulated as a mixed integer nonlinear programming (MINLP)-type problem and solved in General Algebraic Modeling System (GAMS) [5–7]. Moncada et al. (2013) [6] performed an optimization procedure using GAMS in order to select the most promising process pathway for the biorefinery production of fuel ethanol, PHB, and electricity. The selection of the process pathways consisted in the identification of a set of possible routes to transform the feedstocks into products using different criteria such as conversion efficiency and energy consumption. The results from the optimization procedure serve as the criteria for the selection of technologies and the distribution of raw materials.

4.1.1 Assessment Strategies (Scenario-Based Analysis)

Prior to the simulation of biorefineries, the hierarchical decomposition method is applied in order to determine the decomposition of feedstocks and their relationships with products, along with the elements of the onion-based decomposition model [4]. Normally, biorefineries can be divided into different levels of analysis such as batch vs. continuous process, input–output structure of the flow sheet, reaction system synthesis, separation system synthesis, effluent treatment, and heat recovery network. The evaluation of each level of analysis was accomplished with a synthesis tree that allowed the systematic generation and evaluation of the alternatives (scenarios) by the branch and bound method [8,9]. A scenario is defined as a complete process structure that can be used for further evaluation in order to establish whether the proposed configuration fulfills the expectations of the process or if it can be discarded. There are different criteria to define scenarios such as distribution of mass flows of feedstocks/reagents, technologies, and different levels of integration. The selection of the proper criteria to design the scenarios for the biorefinery is based on the objective of the biorefinery. Several authors have used the scenario concept to evaluate different configurations of biorefineries in order to select the best structure network that addresses the biorefinery objective. As an example of the implementation of scenarios in biorefineries, Aristizábal et al. (2015) [10] evaluated the production of ethanol, furfural and hydroxylmethylfurfural (HMF) (furan-based compounds), and octane and nonane (alkanes) using sugarcane bagasse and coffee cut-stems using the biorefinery concept. The authors proposed five scenarios as follows: (i) production of ethanol from hexoses and octane; (ii) production of ethanol, octane, and nonane; (iii) production of the same products of scenario ii, but ethanol and nonane from 40% and 60% of glucose, respectively; (iv) production of ethanol from pentoses and nonane; and (v) production of furfural and HMF. These scenarios were proposed in order to select the best configuration that fulfills the economic and environmental criterion. Figure 4.1 presents the schematic description of the different scenarios that were evaluated by Aristizábal et al. (2015) [10] in coffee cut-stems biorefinery.

As mentioned previously, the main goal of the scenario-based analysis consists in the selection of the process alternative with the best evaluation criterion within each decision level, with the subsequent assessment of the alternatives at the next level of the hierarchical decomposition. The definition of a base case along with other alternatives (scenarios) at each design level is required. As the main result from this assessment, the most promising alternatives are chosen and the ones showing a least favorable performance are rejected [4].

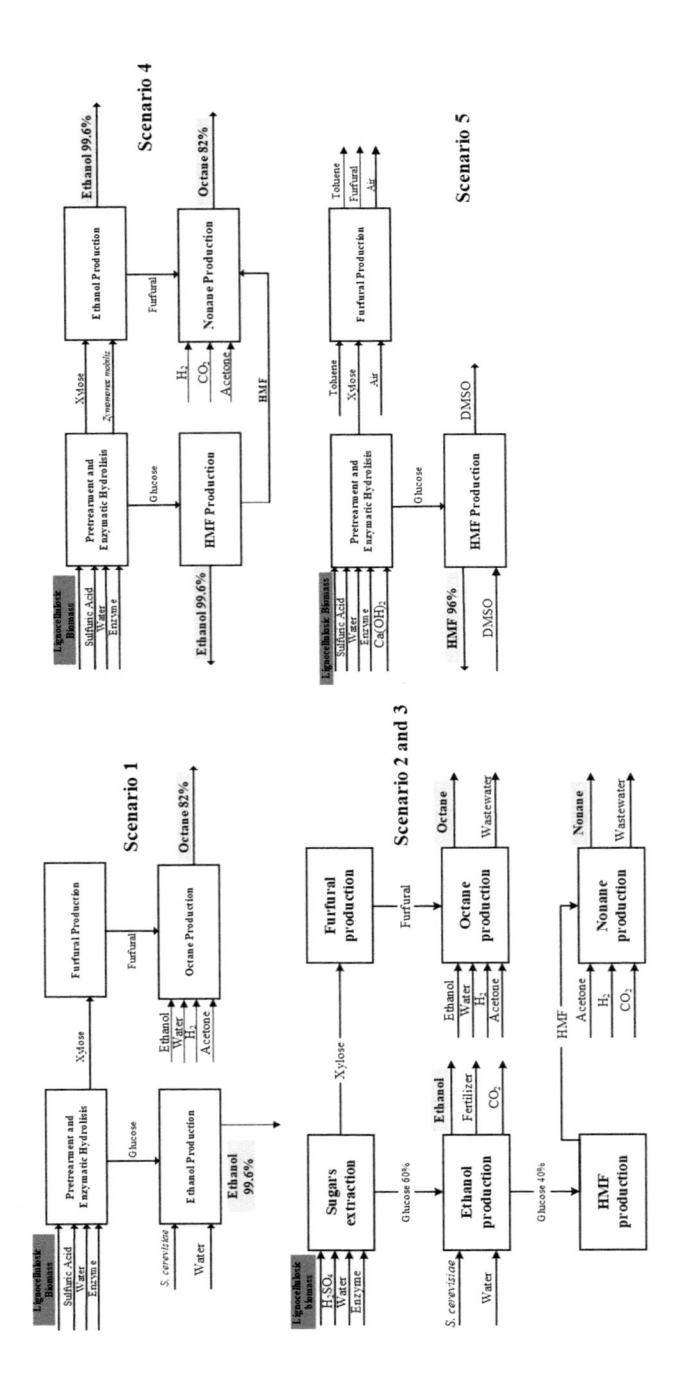

FIGURE 4.1

Schematic description of the scenarios evaluated in the coffee cut-stem biorefinery. (Adapted from Aristizábal et al. (2015) [10].)

4.2 Design Aspects: Problem Definition, Selection of Process Pathways, Biorefining Alternatives and Mass Integration

The design of a biorefinery process is a challenging task due to several factors such as the different types of biomass feedstock and many alternative conversion technologies that can be selected to obtain a wide range of products; therefore, a large number of potential processing paths are available for biorefinery development [11]. Since most of the biorefineries are intended to use natural resources (mainly biomass), the economic and environmental viability of these processes is deeply dependent on local factors such as weather conditions, availability of raw materials, and national or regional subsidies and regulations. Therefore, the design of a biorefinery requires a detailed screening between the set of potential configurations in order to identify the most convenient alternative that fulfills the economic, energy, and environmental goals of the process.

In this sense, a well-defined definition of the problem scope must be stated in order to synthetize and design the proper biorefinery network. The scope of the biorefinery must be specified based on the following items [12].

- Design of the biorefinery network in order to transform raw materials of low added value, into a wide range of products of high added value, and residues, considering operational costs (OPEX) and capital investment (CAPEX).
- Selection of suitable objective functions that allow the determination of the proper biorefinery alternative (based on raw material, products, by-products, and technologies) in order to maximize the conversion efficiency and the process, sustainability, and economic indicators (i.e., products yields, profitability, environmental impacts, etc.).

When the scope of the biorefinery is well defined, the next step involves the design of a superstructure that represents different biorefinery networks (i.e., processing routes to transform biomass into different platforms or products). Each processing route can involve several unit operations that incorporate generic models that describe their operation inside the biorefinery network. The best configuration for the biorefinery is selected based on the sequencing approach previously explained in Chapter 3. The data required for these generic models must be collected from different sources (i.e., literature reports, experimental trials, data bases, etc.) in order to carry out a detailed simulation of the processes involved in the biorefinery network.

4.2.1 Example of the Biorefinery Design Network: Sugarcane Biorefinery

Several authors have used the approach explained in Section 4.2 for the design of different biorefinery configurations [13,14]. Moncada et al. (2013) [6] used this framework in the design of a biorefinery based on sugarcane bagasse to produce three main products: ethanol, polyhydroxybutyrate (PHB), and energy. The selection of the scenarios for the biorefinery was performed based on the description in Section 4.1. In the example of the sugarcane biorefinery described by Moncada et al. (2013) [6], three scenarios were evaluated for technical and economic feasibility of the biorefinery. The scenarios were selected as follows: (i) to minimize the energy consumption, (ii) to maximize the economic margin/potential, and (iii) to minimize the environmental impact.

Many different technologies can be used to obtain the selected products from sugarcane bagasse. However, the biorefinery included key technologies that previously were analyzed in different works [15–20]. The structure of a sugarcane bagasse biorefinery for the integrated production of PHB, ethanol, and energy is shown in Figure 4.2. This figure includes the main components of sugarcane bagasse and platforms, which are raw materials for other processes. The used technologies are numbered with their corresponding name (see, Figure 4.2). At this point, it is very important to note that the split of the main components of sugarcane bagasse is the result of a technological sequence.

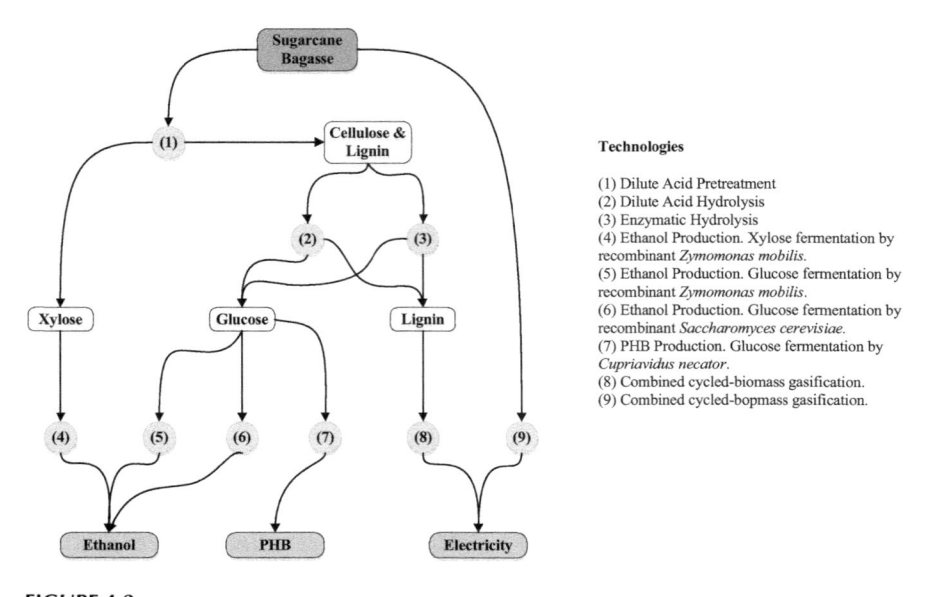

Technologies

(1) Dilute Acid Pretreatment
(2) Dilute Acid Hydrolysis
(3) Enzymatic Hydrolysis
(4) Ethanol Production. Xylose fermentation by recombinant *Zymomonas mobilis*.
(5) Ethanol Production. Glucose fermentation by recombinant *Zymomonas mobilis*.
(6) Ethanol Production. Glucose fermentation by recombinant *Saccharomyces cerevisiae*.
(7) PHB Production. Glucose fermentation by *Cupriavidus necator*.
(8) Combined cycled-biomass gasification.
(9) Combined cycled-bopmass gasification.

FIGURE 4.2
Process pathways for the production of ethanol, PHB, and energy using sugarcane bagasse as a raw material. (Taken from Moncada et al. (2013) [6].)

The first step is the pretreatment, which generally fractionates hemicellulose into pentose (e.g., xylose). Then, the solid fraction (mainly cellulose and lignin) that is not converted into pentose in the pretreatment stage is sent to the cellulose hydrolysis section. The cellulose hydrolysis results in solid fraction (mainly lignin) and liquid fraction (e.g., hexoses). Finally, xylose, glucose, and lignin are used as raw materials for the transformation into the final products.

The necessary data to be fed in the sugarcane bagasse biorefinery design are basically summarized in processing yields and energy requirements for the different technologies. In order to evaluate the technologies presented in Figure 4.2, kinetic models and individual technology simulations are used as the basis for the calculation of the required data. For instance, the kinetic model used for the pretreatment and dilute acid cellulose hydrolysis steps was reported by Jin et al. (2011) [15]. The kinetic model for enzymatic cellulose hydrolysis was reported by Morales-Rodriguez et al. (2011) [16]. The kinetic model for ethanol production using recombinant *Zymomonas mobilis* was reported by Leksawasdi et al. (2001) [19]. The kinetic model used for the calculation of PHB production was reported by Shahhosseini (2004) [18]. The biomass gasification coupled with a combined cycle was simulated using the Aspen Plus software. This procedure was done for technologies (8) and (9) (see Figure 4.2) using lignin and sugarcane bagasse, respectively. It is very important to mention that the calculation of energy requirements was done using the Aspen Plus software for each technology. These requirements include different equipment such as heat exchangers which are important in the adequacy of upstreams and downstreams of each technology.

Once the yields (mass and energy) were determined, the mass and energy balances were used in the formulation of the optimization problem. Table 4.1 summarizes the yields of the main intermediates/products for each of the proposed technology in the sugarcane bagasse biorefinery (see the numbers in Figure 4.2). On the other hand, each technological pathway has its own energy requirements; therefore, a cost is related to each one. These energy costs refer to the utilities needed to cover the supply. Therefore, the results of the energy requirements were used to evaluate the economic and energy performance of the biorefinery. Table 4.2 shows the energy requirements per unit of product or platform for each of the technologies.

4.2.2 Mass Integration

After the design of the different biorefinery alternatives (scenarios), different integration approaches can be applied in order to maximize the use of the resources in the process. Integration can be performed between raw materials, technologies, products, and even energy requirements (hot and cold utilities). The idea of this approach is to generate and select the biorefinery network that generates no wastes and low-energy requirements for external sources (self-sufficient). El-Halwagi (2012) [21] performed a detailed study of the mass integration framework for biorefineries using big-picture

TABLE 4.1

Yields of the Products from Each Technology Involved in the Sugarcane Biorefinery Described by Moncada et al. (2013) [6]

Technology[a]	Yield	Units
1	0.74	g xylose/g hemicellulose
2	0.89	g glucose/g cellulose
3	0.75	g glucose/g cellulose
4	0.48	g ethanol/g xylose
5	0.47	g ethanol/g glucose
6	0.49	g ethanol/g glucose
7	0.31	g PHB/g glucose
8	5108.78	MJ/ton dry lignin
9	5696.94	MJ/ton dry bagasse

[a] The numbers describe each technology used in Figure 4.1.

TABLE 4.2

Energy Needs and Energy Costs of Each Technology Involved in the Sugarcane Biorefinery Described by Moncada et al. (2013) [6]

| Technology[a] | Energy Needs | | | Energy Costs (USD/MJ) | |
	Heating	Cooling	Units (MJ/ton)	Heating	Cooling
1	8537.12	0.00	Xylose	6.30×10^{-4}	7.85×10^{-5}
2	4389.06	2936.25	Glucose	6.30×10^{-4}	7.85×10^{-5}
3	2194.53	587.25	Glucose	6.30×10^{-4}	7.85×10^{-5}
4	143238.45	57306.17	Ethanol	7.68×10^{-4}	7.85×10^{-5}
5	71949.45	28785.20	Ethanol	7.68×10^{-4}	7.85×10^{-5}
6	69012.74	27610.29	Ethanol	7.68×10^{-4}	7.85×10^{-5}
7	44010.42	27184.93	PHB	6.30×10^{-4}	7.85×10^{-5}

[a] The numbers describe each technology used in Figure 4.1.

approaches through different cases: minimum waste discharge, fresh usage, and product yield. According to this study, the mass integration is a systematic methodology that allows identifying and understanding the global mass flow within the process, provides a fundamental understanding of the performance targets, and allows optimizing the generation and routing of species throughout the process [21]. The application of the mass integration strategies does not require a prior specific change in the process in order to identify the mass targets. The main goal of these strategies is to develop cost-effective modifications of the process that can include stream segregation/mixing, recycle, interception using separation devices, changes in design and operating conditions of units, materials substitution, and technology changes including the use of alternate chemical pathways [21].

Several authors have used the mass integration concept in order to evaluate the performance of different biorefinery configurations [22,23]. An example of the implementation of the mass integration concept was described by Moncada et al. (2015) [24] in the design of a biorefinery using castorbean and microalgae as raw materials to produce polyol, ethylene glycol, omega-3 acid, biodiesel, and methanol. Three scenarios were proposed to evaluate the technical, economic, and environmental feasibility of the biorefinery network and the effect of different integration levels (mass and energy) in the design of the process. The first scenario only considers internal mass integration of process streams (e.g., methanol, glycerol) and also assumes that CO_2 for microalgae comes from an external source (no cogeneration included). The second scenario describes a system in which internal mass integration, as well as external mass integration (water sources), and heat integration (pinch) are considered. The third scenario considers mass and heat integration as well as energy supply by the cogeneration system. A graphic description of the scenarios involved in the biorefinery design is presented in Figure 4.3.

The results suggested that different levels of integration should be considered to improve the overall efficiency of multiproduct integrated portfolios. It was also evident that the energy and mass integration have an important effect on the economic and environmental performances of biorefinery systems. From the case studied, the scenario with the highest integration level (mass+energy integration) showed the best economic and environmental performance. Consequently, this case is a very good example to demonstrate how the mass integration concept can be really successful in the biorefinery design.

4.3 Technology Integration

Technology integration is a process performed aiming to increase the whole efficiency of a biorefinery system in terms of raw materials and energy requirements. Thus, technology integration involves the analysis of the material and energy streams in the process as well as the processing lines employed in the biorefinery. These analyses can be carried out through mass and energy integration. Nevertheless, other forms to improve the overall behavior of a biorefinery system is through the combination of two or more technologies in the same process equipment as is the case of the simultaneous saccharification and fermentation, reactive distillation, extractive fermentation, and extractive dividing wall column [25–27]. Moreover, the energy performance of a biorefinery can be improved through the implementation of pinch analysis, the inclusion of a biomass integrated gasification combined cycle (BIGCC), a cogeneration plant, and a trigeneration facility. These options can contribute to improve the economic indicators of the biorefinery decreasing

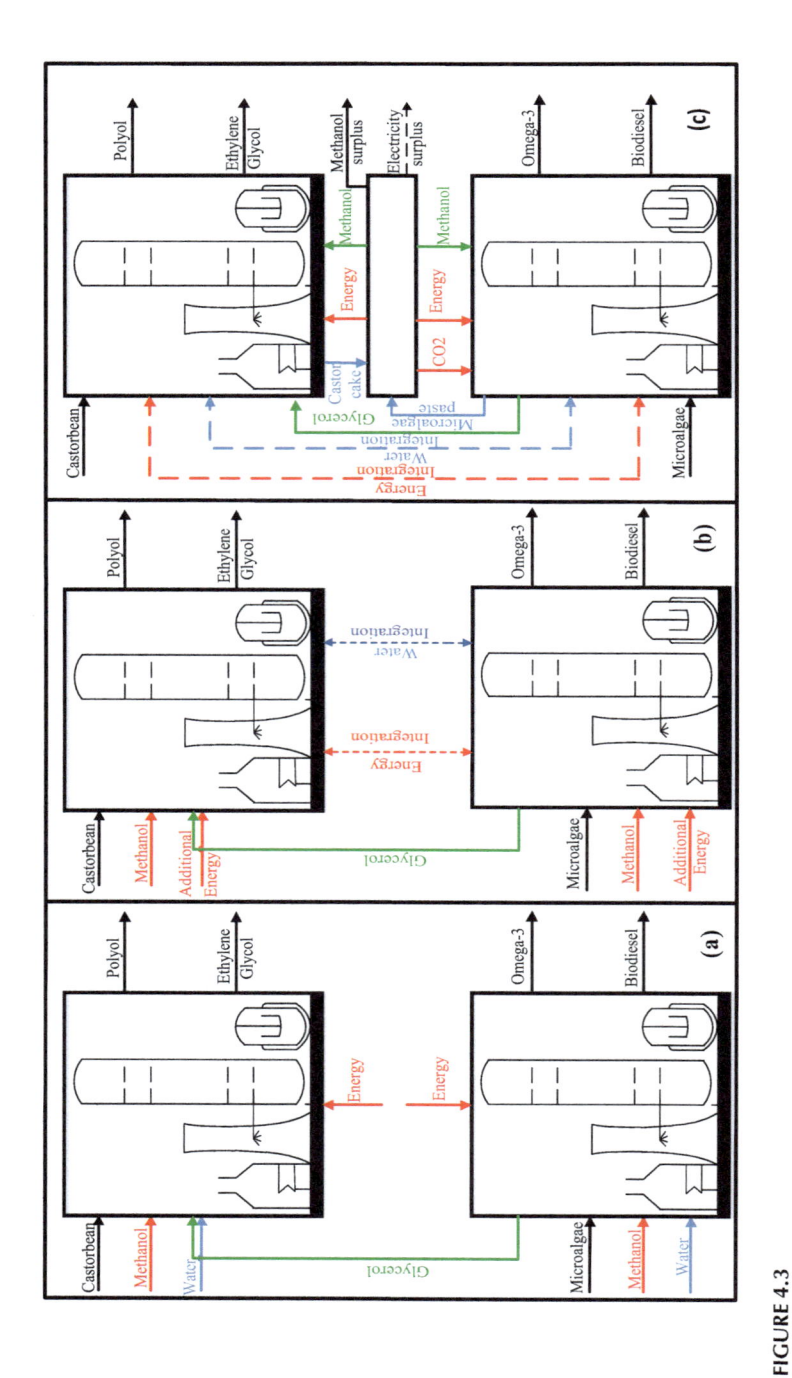

FIGURE 4.3
Description of the implementation of the mass integration concept in the castor bean and microalgae biorefinery described by Moncada et al. (2013) [24].

the associated costs with the production of utilities (i.e., steam and power) as well as to decrease the potential environmental impact (PEI) [28,29].

In this way, the mass and energy analysis plays an important role in the design and optimization of biorefinery systems. Nevertheless, an in-depth description of the energy analysis considering the energy savings that can be produced by the implementation of cogeneration systems, reaction–separation systems, and heat integration will be discussed and explained in the Subsection 4.3.1 through the analysis of different case studies reported in the literature. On the other hand, the mass integration savings and improvements contributed by this type of analysis have been already analyzed in the Subsection 4.2.2. Therefore, this will not be analyzed in the following text.

4.3.1 Energy Consumption, Integration, and Cogeneration

The energy analysis of a biorefinery system must be performed once the material balance and the process conditions have been defined. Thus, this analysis can be accomplished taking into account a global perspective or a more detailed and focused methodology. The first option is to realize a global energy balance with the inputs and outputs of the process. The input is the energy content of the raw materials, whereas the output is the energy content of the obtained products as well as the overall heat and power requirements of the biorefinery [30]. Nevertheless, this type of analysis can be used only to describe the overall performance of the process and can also be used for comparison purposes. The second option (i.e., step-by-step analysis) gives more detailed information about the possibilities to improve the process through the energy integration, the implementation of utilities production system and simultaneous processes. Therefore, a detailed energy analysis of a biorefinery system should be done following the next steps.

- *Energy consumption analysis:* This step is related with the identification of the main processes in a biorefinery that have the highest energy requirements in terms of heat and power. The energy consumption analysis consists in finding the mass flows of low-, middle-, and high-pressure steam that are necessary to supply the overall heat requirements of the process. On the other hand, the electrical needs related with the use of pumps, compressors and mixers must be taken into account also to complete the energy analysis of a biorefinery.

In a biorefinery system, these needs can be evaluated using the Aspen Energy Analysis tool to calculate the amount and quality of the necessary steam to carry out the process. However, if this tool is not available, it is possible to calculate this data through the application of heat transfer concepts and the energy balances of each unit into the biorefinery. Moreover, the energy consumption analysis involves identifying the main process areas (e.g., pretreatment, saccharification, reaction, separation, purification) that have the highest energy needs in terms of heat and power. Additionally, this analysis should be used to classify hot and cold streams aiming to perform the composite curves as the initial phase of the pinch analysis.

A techno-economic and energy evaluation of different schemes of biorefineries was performed by García et al. (2017) [31]. In general, three scenarios were evaluated by these authors. The first scenario was the stand-alone production of hydrogen through the air-blown downdraft gasification of coffee cut-stems. The second scenario was the hydrogen obtainment and the electricity generation. Finally, the third scenario was the hydrogen, electricity, and ethanol production. These schemes were proposed and designed following the hierarchy, sequencing, and integration concepts described in Chapter 3, and they are shown in Figure 4.4.

García et al. (2017) [31] describes the energy analysis of the three scenarios mentioned previously. First, the energy efficiency of the process, which is defined as the ratio between the energy produced by the products (e.g., hydrogen and ethanol) and the energy released by the feedstock, was calculated. Second, the net energy value (NEV) was calculated. This indicator is related with the amount of energy that is required in the overall production process over the energy generated by the main products of the biorefinery.

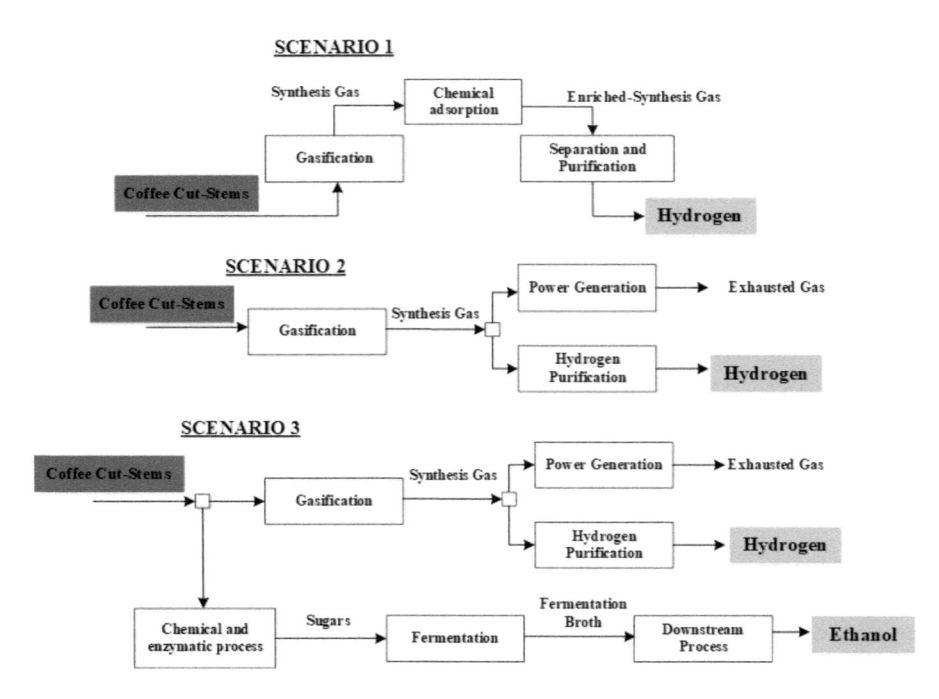

FIGURE 4.4
Block diagram of the scenarios evaluated by García et al. (2017) [31].

According to the data reported by these authors, the energy efficiency of each scenario was 62.06%, 58.20%, and 94.50%, respectively. This result implies that the integration of other products in the hydrogen production such as heat, power, and ethanol increases the energy efficiency of the whole process in comparison with the stand-alone production of hydrogen. On the other hand, the NEV of these scenarios also is showed by these authors. In accordance with the energy efficiency results, scenario 3 showed the best NEV of the three scenarios. Therefore, it was possible to conclude that scenario 3 shows the best energy performance of the three proposed scenarios. Moreover, in this case study, a graphical representation of the global energy balance of each scenario is shown. This graphical representation is called as Sankey diagram, which presents the main inputs and outputs of a process with arrows. This diagram is very interesting because it shows the highest values in bigger arrows than the low values. The Sankey diagrams for each proposed scenario by García et al. (2017) [31] are presented in Figure 4.5.

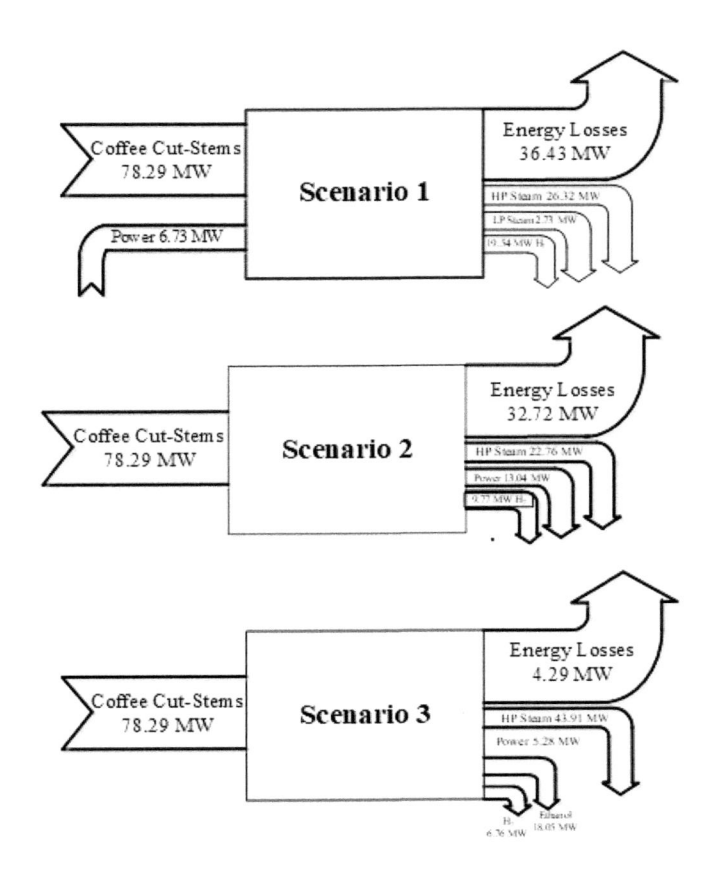

FIGURE 4.5
Sankey diagrams for the scenarios proposed by García et al. (2017) [31].

The Sankey diagram for scenario 1 shows that about 36.43 MW of energy is lost in the stand-alone hydrogen production process. These energy losses are decreased in the other two scenarios. In contrast, the amount of high- and low-pressure steam as well as the net power outputs of the processes increases. This reflects that the implementation of hierarchy, sequencing, and integration concepts for the biorefinery system design strongly impacts the energy performance of a process.

Another example related with the energy analysis of biorefinery systems was presented by Cardona et al. [32]. In this case study, a biorefinery for the production of ethanol, xylitol, biogas, and lactic acid was analyzed. From the energy analysis point of view, the authors reported the energy requirements for each stage of the process. This analysis was performed aiming to identify the main areas of the biorefinery system to be improved from the energy point of view. The process block diagram of the biorefinery analyzed by Cardona et al. (2018) [32] is shown in Figure 4.6.

The results reported by these authors showed that the main energy-consuming areas in this biorefinery were the xylitol and lactic acid production processes; meanwhile, the pretreatment and enzymatic saccharification processes were the areas with the lowest energy consumption. These results are presented in Figure 4.7.

In accordance with the energy consumption shares reported in Figure 4.7, the lactic acid and xylitol production process must be analyzed first than the other blocks of the biorefinery aiming to identify the main parts of the process that can be improved to decrease, if it is possible, the energy demand of the

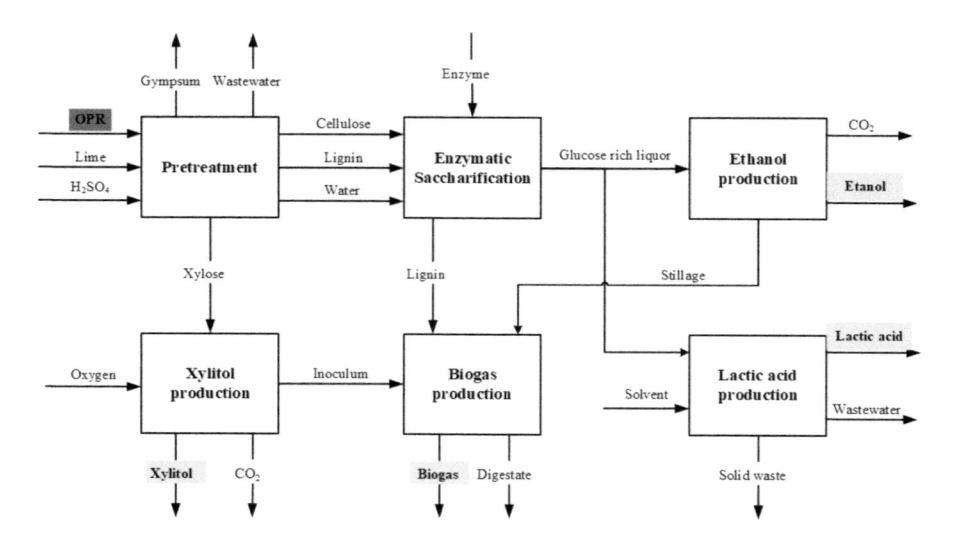

FIGURE 4.6
Block diagram of a biorefinery for ethanol, lactic acid, biogas, and xylitol production proposed by Cardona et al. (2018) [32]. OPR: oil palm rachis.

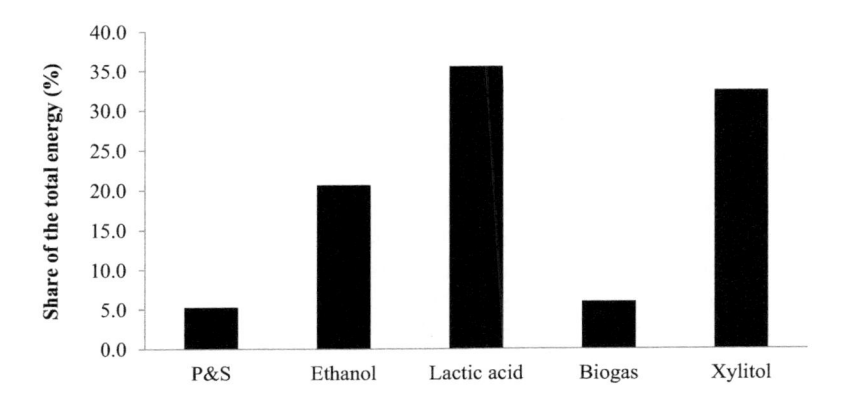

FIGURE 4.7
Share of energy consumption of the biorefinery proposed by Cardona et al. (2018) [32].

process. Moreover, these energy analysis results suggest that the implementation of other type of technologies such as reaction–separation processes could be done to reduce the energy consumption, even the implementation of a cogeneration system using the main residues of the processing lines of the biorefinery for heat and power generation. Finally, once the energy analysis has been performed, the main indicators of the process have been calculated (e.g., energy efficiency), and the main areas that consume a great amount of energy have been identified, the next step is to perform an energy integration using pinch analysis or another methodology reported in the literature.

- *Energy integration:* This step aims to reduce the energy requirements of the biorefinery through the use of the energy of hot and cold streams. For this, heat exchanger networks are used as the main tool to propose an energy-integrated system. The first stage in the heat exchanger network system design is to perform the composite curves of the process, which are composed by the overall hot and cold streams in the biorefinery. Then, a pinch analysis must be performed to find the maximum energy recovery of the process minimizing the capital costs associated with the implementation of heat exchangers. Finally, the implementation of the heat exchanger network is performed [33].

The energy integration process together with mass integration, generally, has a good effect on the economic and environmental indicators of a biorefinery. This fact could be appreciated in the biorefinery designed by Moncada et al. (2015) [24] where the energy savings after implementing the energy integration concept were about 40%. Furthermore, pinch analysis can be used to improve the economic margin of a product in a biorefinery. This is the case presented by Dias et al. (2009) [34], where the pinch analysis was used to determine the minimum hot utility needed in an ethanol production process in a sugarcane biorefinery. In addition, using the pinch methodology, these authors calculated the maximum availability of bagasse that can be used for ethanol production and energy production to self-supply its energy demands.

- *Cogeneration:* This process can be defined as a thermodynamically efficient way of using energy, which is able to cover complete or partially both heat and electricity requirements of a factory [35]. There are a lot of technologies used for cogeneration. However, one of the most employed technologies in biorefineries is the BIGCC, which is shown in Figure 4.8. The basic elements of the BIGCC system include biomass dryer, gasification chamber, gas turbine, and heat recovery steam generator (HRSG). Gasification is a thermochemical conversion technology of carbonaceous materials (coal, petroleum coke, and biomass), to produce a mixture of gaseous products (CO, CO_2, H_2O, H_2, CH_4) known as syngas added to small amounts of char and ash. Gasification temperatures range between 675°C and 1000°C [36]. The gas properties and composition of syngas change according to the used gasifying agent (air, steam, steam–oxygen, oxygen-enriched air), gasification process, and biomass properties [37]. A gas turbine is a rotator engine that extracts energy from a flow combustion gas. It is able to produce power with an acceptable electrical efficiency, low emission, and high reliability [38]. A gas turbine is composed of three main sections: compression (air pressure is increased, aimed to improve combustion efficiency), combustion (adiabatic reaction of air and fuel to convert chemical energy to heat), and expansion (obtained pressurized hot gas at high speed passing

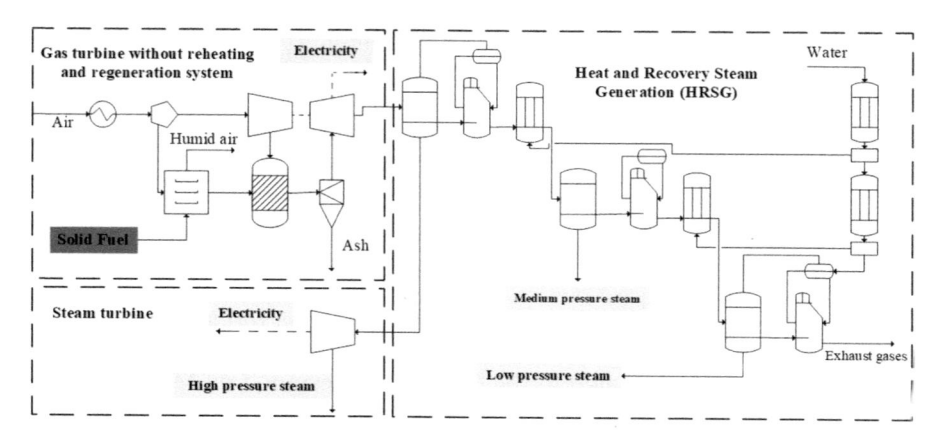

FIGURE 4.8
BIGCC system described by Quintero et al. (2013) [36].

through a turbine generating mechanical work) [17]. An HRSG is a high-efficiency steam boiler that uses hot gases from a gas turbine or reciprocates engine to generate steam, in a thermodynamic Rankine cycle [17]. The system is able to generate steam at different pressure levels. In general terms, the solid residues generated in a biorefinery are used in a cogeneration system to cover at least part of the energy requirements.

The cogeneration process showed in Figure 4.8 has a gas turbine without a reheating and regeneration system. However, this type of turbines is not mostly used due to their low thermal efficiency. Therefore, improvements on this design can be done. In addition, a co-firing system is added aiming to increase the amount of steam that can be produced in the HRSG system as well as the amount of power derived from the steam turbine. This design is presented in Figure 4.9.

4.3.2 Reaction–Separation Processes

Reaction–separation processes are those in which two processes are coupled through thermodynamic concepts to improve the conversion of a desired product. These processes have many advantages with respect to conventional processes. Some of these advantages are high productivities and yields, low

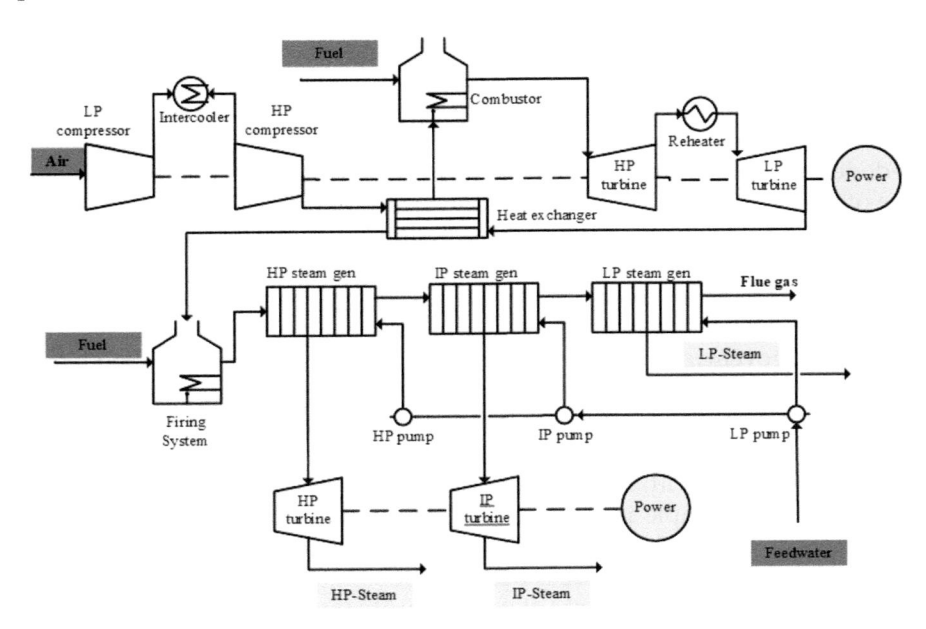

FIGURE 4.9
Cogeneration system with an intercooling, reheating, and regenerative cycle and co-firing system [25].

capital investment costs, lower energy requirements, easy recovery of the desired products, and lower space demand into the biorefinery. Therefore, this type of processes has been researched in the past few years. The most common reaction–separation processes that are found at industrial level are reactive distillation, reactive liquid extraction, and extractive fermentation.

One of the abovementioned options to perform the reaction–separation process in a biorefinery is the extractive fermentation. This process arises as an option to overcome the inhibition effect that the end products have on either cell growth rate or product biosynthesis rate as it occurs in the production of acetone and butanol, lactic acid, and ethanol. This process improves the fermentation yields and performance through a biocompatible extracting agent (solvent), in such a way that the product migrates to the solvent. Therefore, the extractive fermentation combines the liquid–liquid extraction principles with the fermentation process. This process has been studied by several authors because it can improve both batch and continuous processes reducing capital, operating, and maintenance costs and increasing the productivities of the overall process. In the following text, two examples of this coupled process are analyzed from an energy point of view aiming to understand the impact that such processes have on a whole biorefinery system.

Gutierrez et al. (2013) [39] studied the extractive fermentation process in batch and continuous modes for ethanol production using pentoses and hexoses as substrates. These authors simulated the extractive fermentation process using n-dodecanol as a solvent. They found that the process productivity, using 400 g/L of fermentable sugars as the initial condition, was 1.89 g/L/h until the substrate concentration is exhausted. This value is about 30% higher than the productivity when the fermentation is performed without any solvent. Moreover, these authors report that the addition of n-dodecanol reduces the time of the fermentation process to about 50%. On the other hand, the continuous operation is highly affected by the presence of an immiscible solvent. In fact, the productivity of the process increases three times in comparison with the process without solvent. These results imply that the energy consumption of the fermentation process and the ethanol recovery process will be lower in comparison with the conventional process. The abovementioned can be explained because lower operation times require low amount of steam to maintain the temperature of the process. Moreover, the ethanol–n-dodecanol is a binary mixture that does not have a binary azeotrope. Therefore, a simple distillation process can be carried out to recover ethanol with a high-volume fraction (>98%).

The second example is related with the extractive fermentation used to produce lactic acid. This case study was presented by Wasewar et al. (2004) [40]. In this case study, glucose is used as a substrate to be converted into lactic acid. These authors report different solvents that may be used to remove the lactic acid from the broth. In addition, they give the main reasons for which the extractive fermentation process overcomes all energy problems of the conventional lactic acid recovery process. Among the main reasons, it is possible to find the high capital investment savings that the extractive fermentation process gives to the overall economic indicators of the productive process. Therefore, the lactic acid case can be considered as one of the most visible examples that demonstrated the applicability of the extractive fermentation process. Finally, there are a lot of reaction–separation processes that can be implemented in a biorefinery system such as the reactive distillation column and the packed-bed enzyme saccharification and fermentation process [41–43].

4.4 Techno-Economic Indicators

Economic analysis considers capital and operating costs, which can be estimated based on the design information generated in process modeling. Figure 4.10 presents a descriptive guideline of the different stages that are required in order to evaluate the economic performance of biorefineries. The starting point of the techno-economic analysis is the mass and energy balances from the simulation procedure that are used for two purposes: the sizing of the operation units (equipment) of the simulated biorefinery and the determination of the energy requirements. The equipment sizing is carried out in order to determine the costs involved in the acquisition of the equipment, and these costs are estimated in the complementary software Aspen Process Economic Analyzer (Aspen Technology Inc.). On the other hand, the energy requirements (heating and cooling utilities) are calculated from the Aspen Energy Analyzer (Aspen Technology Inc.) using the pinch methodology previously explained in Section 4.3. The amount of heating and cooling

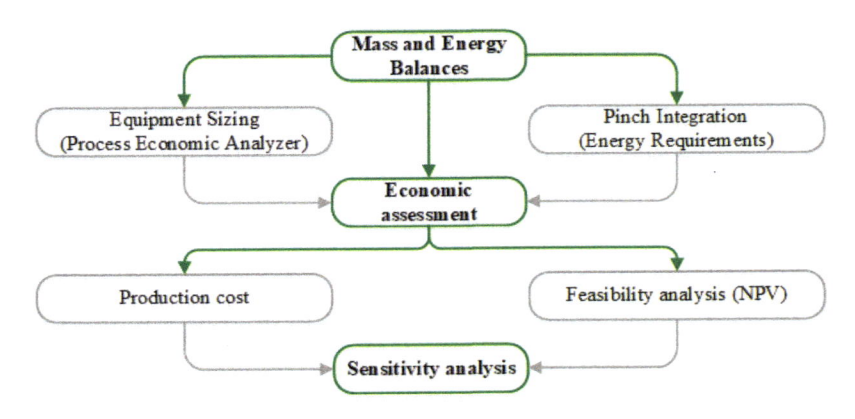

FIGURE 4.10
Descriptive guideline of the techno-economic analysis of biorefineries.

requirements is used to calculate the costs associated with utilities in the biorefinery.

Subsequently, the economic analysis of the biorefinery is carried out in order to determine the total capital investment (TCI) that includes the funds required to purchase land, design, purchase, and install equipment and buildings, as well as to bring the facility into operation. Some of the items that are required to determine the TCI are land, fixed capital investment, offsite capital, allocated capital, and working capital, among others. The calculation of these costs is based on the estimation methods proposed by Peters and Timmerhaus (2004) [44]. The fixed capital investment is estimated based on purchased equipment costs and typical factors to account for delivery (10% of purchased equipment costs), direct costs (percentage of purchased equipment costs: installation 47%, instrumentation and control 36%, piping 68%, electrical systems 11%, buildings 18%, yard improvements 10%, service facilities 70%), and indirect costs (percentage of purchased equipment costs: engineering and supervision 33%, construction expenses 41%, legal expenses 4%, contractor's fee 22%, contingency 44%) [44]. These costs must be constantly updated to the current market prices using the Chemical Engineering Plant Cost Index.

Based on the estimation methods proposed by Peters and Timmerhaus (2004), the annualized costs considered features such as raw materials, utilities, maintenance, labor, fixed and general, overhead, and capital depreciation [44]. Raw materials costs are estimated based on consumption levels from mass balances and market prices. Utilities costs are estimated based on energy balances [45]. Maintenance costs consist of maintenance cost (6% of fix capital investment) and operating supplies cost (15% of maintenance) [44]. Labor costs consist of operating labor cost (three shifts of 8 h each, 10 operators per shift), operating supervision cost (15% of operating labor) and laboratory charges cost (10% of operating charges) [44]. Fixed and general costs involve taxes (2% of fix capital investment), insurance (1% of fix capital investment), and general cost (20% of labor, supervision, and maintenance) [44]. Plant overhead is estimated as 60% of labor, supervision, and maintenance [44]. Annual capital depreciation is estimated based on the fix capital investment and the start-up material cost (10% of fix capital investment), which are related to the interest rate (depending on the country context) and a lifetime of 10-years.

4.4.1 Allocation Factors

Initially, the cost estimation for each scenario can be performed for the whole biorefinery system, but it should be noted that each scenario is a multiproduct portfolio. Therefore, to calculate the individual production costs per product, it is necessary to assign individual factors (allocation factors) to the total costs of the entire biorefinery. There are many ways to allocate costs in multiproduct portfolios (mass, energy, economic, environmental, etc.); however, the most widely studied are the mass and economic allocation methods. The mass allocation method uses the distribution of the product flow rates in order to "allocate" the production costs based on the productivity of the biorefinery. If energy-driven and product-driven biorefineries are considered in the same scenario, this allocation method does not address correctly the distribution of the production costs because different measure units (energy-driven [MJ/h] and product-driven [kg/h] biorefinery) are used. On the other hand, the economic allocation ensures a fairer assessment of the product values (prices) and avoids over assignment of the overall process cost to a low-value co-product [46]. The allocation factors can be calculated based on the contribution to sales of each product. After allocating the annualized costs to each product, these are normalized by their respective production volume [46].

Both allocation methods have been used to evaluate the production costs of the biorefinery and its distribution among the main products. The mass allocation was previously used by Hernández et al. (2014) [47] in the determination of the production costs of ethanol, xylitol, furfural, and PHB using olive stones as a raw material. Table 4.3 presents the summary of the economic results from the olive stone biorefinery considering the mass allocation method. As can be evidenced, more than 60% of the production costs are distributed between the production of ethanol and that of xylitol because of the high production volume of these two products, despite the fact that these products have a low added value (market price).

On the other hand, the economic allocation method was used by Moncada et al. (2016) [48] to determine the production costs of ethanol and furfural in a *Pinus patula* bark biorefinery. Table 4.4 presents the summary of the production costs of the proposed biorefinery using the economic allocation method. Despite the high electricity generation from the biorefinery, the economic allocation methods allow distributing the production costs that generate the highest profits from the process. It is noteworthy that more than 90% of the production costs are allocated to the ethanol and furfural production, which agrees with the complex equipment and energy requirements.

TABLE 4.3

Production Costs of the Olive Stone Biorefinery Using the Mass Allocation Method

Products	Productivity (ton/day)	Mass Allocation (%)	Cost Distribution (Million USD/year)
Xylitol	12	31	18.83
Furfural	8.04	21	12.62
Ethanol	14.36	37	22.53
PHB	4.07	11	6.39
Total	**38.47**	**100**	**60.36**

TABLE 4.4

Production Costs of the *P. patula* Bark Biorefinery Using the Economic Allocation Method

Products	Productivity	Market Price	Revenues (Million USD/year)	Economic Allocation (%)
Ethanol	4,080.8 kg/h	1.34 (USD/kg)	47.93	39
Furfural	6,485.3 kg/h	1.20 (USD/kg)	68.22	56
Electricity	22,130 MJ/h	0.028 (USD/MJ)	5.39	4

4.4.2 Feasibility Analysis

To assess the profitability of the biorefinery system, the net present value (NPV) can be used as an economic parameter. This parameter indicates the profits/gains of the process during the project life, which in chemical processes is between 10 and 15 years. This assessment also considers important data such as inflation rate of production costs, income tax, and minimum acceptable rate of return. Table 4.5 shows some of the market prices normally used in the economic assessment of different biorefinery networks. Additionally, another parameter that could influence the profitability of the biorefinery is the payback period, which can be defined as the time required

TABLE 4.5

Example of Prices Used in the Economic Evaluation of the Biorefinery

Item	Price	Unit
Biomass[a]	40	USD/ton
Natural gas[b]	2.84	USD/GJ
Electricity[b]	0.1	USD/kWh
Sulfuric acid[c]	94	USD/ton
Ethanol[b]	1.06	USD/L
Sodium hydroxide[d]	98	USD/ton
Enzyme[d]	700	USD/ton
Process water[e]	0.07	USD/m^3
Low P. steam (3 bar)[e]	10.84	USD/ton
Operating labor[b]	4.5	USD/h
Furfural[d]	1200	USD/ton

Type of truck: 10 ton truck. Diesel price: 4.11 USD/gallon.

[a] Price including transportation charges. Average traveled distance: 100 km.

[b] Colombian national average.

[c] Prices based on ICIS pricing indicatives (ICIS, 2014).

[d] Prices based on Alibaba international prices (ALIBABA, 2014).

[e] Estimated based on [45] and updated to 2014 using the Chemical Engineering Plant Cost Index.

to recover the initial investment of the project. These two parameters are key criteria in the selection of feasible scenarios in a biorefinery approach. In this sense, García et al. (2017) [31] evaluated the production of hydrogen through different biorefinery scenarios using coffee cut-stems as a raw material. Three scenarios were evaluated as follows: (i) direct production of hydrogen, (ii) distribution of the synthesis gas flow rate for hydrogen and electricity production, and (iii) the same distribution of scenario ii plus a feedstock distribution for ethanol fermentation. Figure 4.11 presents the results from the feasibility analysis of the proposed scenarios. From this analysis, scenario iii has the highest NPV due to the implementation of the biorefinery concept; however, this same behavior was not evidenced in scenario ii that has the lowest NPV of the evaluated scenarios. The payback period of the scenarios evidenced that despite the fact that scenario iii generates the highest profits, it requires a longer time to recover the initial investment in comparison with scenario i. This can be explained because of the high initial investment of the process as evidenced in the first 2 years of investment before the process starts.

4.4.3 Sensibility Analysis

Prices may vary over time, and some of them may be very uncertain (e.g., biomass price, operating costs, enzyme price). A sensitivity analysis can be carried out to assess how the economic parameter affects the NPV the most. Variation on prices considers values up to 100% above and below the reference prices displayed in Table 4.5. Additionally, the effect of plant capacity (raw material processing capacity) can be used as the economic criterion to select the amount of feedstock that must be processed so the biorefinery will be economically feasible. To do this, linear and sixth-tenths rules are considered for scaling operating and capital costs, respectively.

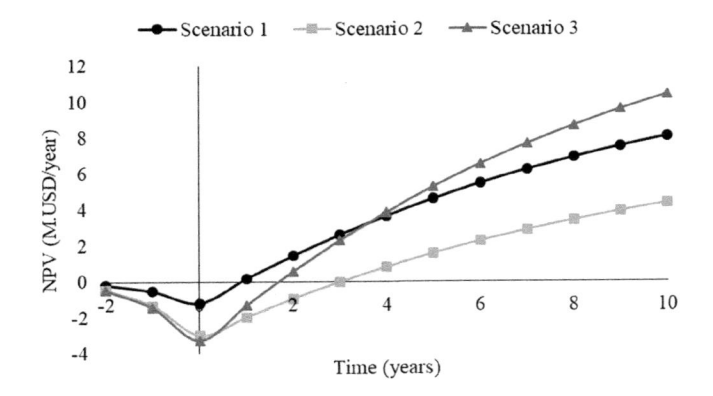

FIGURE 4.11
Feasibility analysis of the biorefinery scenarios in the coffee cut-stem biorefinery reported by García et al. (2017) [31].

Following the example of the coffee cut-stem biorefinery to produce bio-energy proposed by García et al. (2017) [31], the variation of the raw material and product market price was also evaluated in order to determine the fluctuation in the profitability of the proposed scenarios as presented in Figure 4.12. The parameters that have a strong influence in the NPV are the ethanol and hydrogen prices because a reduction of 50% and 90% in the

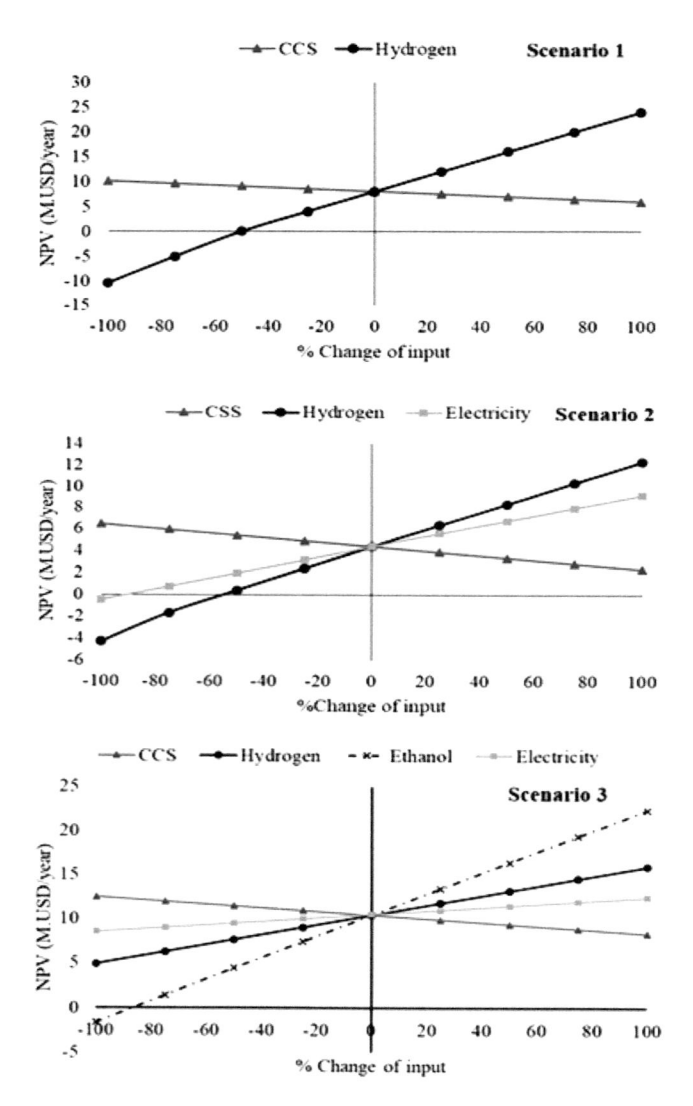

FIGURE 4.12
Sensibility analysis of the effect of the market price variation in the NPV of the biorefinery.

market price, respectively, may turn the process nonprofitable. Coffee cut-stems and electricity market price do not affect considerably the economic performance of the biorefinery. In scenario 1, the hydrogen market price has the strongest influence in the NPV behavior because it is the only product obtained from the stand-alone scheme. The same behavior was evidenced in scenario 2, despite the fact that electricity is also produced. By contrast, in scenario 3, the ethanol market price has the strongest influence in the NPV behavior because of the high production volume and thus its high contribution to the economic allocation of the scenario.

4.5 Conclusion

The use of simulation tools to analyze different biorefinery schemes from a techno-economic and an environmental point of view has great potential in the formulation of processes with high levels of sustainability. However, concepts such as mass and energy integration are key in the development of efficient biorefinery schemes. These processes are only possible if a correct implementation of the concepts of hierarchy and sequencing described in Chapter 3 have been carried out.

References

1. G. Liu, J. Zhang, and J. Bao, "Cost evaluation of cellulase enzyme for industrial-scale cellulosic ethanol production based on rigorous Aspen Plus modeling," *Bioprocess Biosyst. Eng.*, vol. 39, no. 1, pp. 133–140, 2016.
2. C. A. García, M. Morales, J. Quintero, G. Aroca, and C. A. Cardona, "Environmental assessment of hydrogen production based on *Pinus patula* plantations in Colombia," *Energy*, vol. 139, pp. 606–616, 2017.
3. N. Hajjaji, S. Martinez, E. Trably, J. P. Steyer, and A. Helias, "Life cycle assessment of hydrogen production from biogas reforming," *Int. J. Hydrogen Energy*, vol. 41, no. 14, pp. 6064–6075, 2016.
4. Ó. J. Sánchez and C. A. Cardona, "Conceptual design of cost-effective and environmentally-friendly configurations for fuel ethanol production from sugarcane by knowledge-based process synthesis," *Bioresour. Technol.*, vol. 104, pp. 305–314, 2012.
5. P. Cheali, K. V. Gernaey, and G. Sin, "Synthesis and design of optimal biorefinery using an expanded network with thermochemical and biochemical biomass conversion platforms," in *23rd European Symposium on Computer Aided Process Engineering*, 1st ed., A. Kraslawski and I. Turunen, Eds., pp. 985–990. San Diego, CA: Elsevier, 2013.

6. J. Moncada, L. G. Matallana, and C. A. Cardona, "Selection of process pathways for biorefinery design using optimization tools: A Colombian case for conversion of sugarcane bagasse to ethanol, poly-3-hydroxybutyrate (PHB), and energy," *Ind. Eng. Chem. Res.*, vol. 52, no. 11, pp. 4132–4145, 2013.

7. L. E. Rincón, M. J. Valencia, V. Hernández, L. G. Matallana, and C. A. Cardona, "Optimization of the Colombian biodiesel supply chain from oil palm crop based on techno-economical and environmental criteria," *Energy Econ.*, vol. 47, pp. 154–167, 2015.

8. L. T. Biegler, I. E. Grossmann, and A. W. Westerberg, *Systematic Methods of Chemical Process Design*. Upper Saddle River, NJ: Prentice Hall, 1997.

9. W. D. Seider, J. D. Seader, D. R. Lewin and S. Widagdo, *Product & Process Design Principles—Synthesis, Analysis & Evaluation*. Chichester: Wiley, 2008, pp. 1–768.

10. V. Aristizábal, Á. Gómez, and C. A. Cardona, "Biorefineries based on coffee cutstems and sugarcane bagasse: Furan-based compounds and alkanes as interesting products," *Bioresour. Technol.*, vol. 196, pp. 480–489, 2015.

11. D. K. S. Ng, R. R. Tan, D. C. Y. Foo, and M. M. El-Halwagi, *Process Design Strategies for Biomass Conversion Systems*. Wiley, 2015, pp. 1–361.

12. A. Quaglia, B. Sarup, G. Sin, and R. Gani, "Integrated business and engineering framework for synthesis and design of enterprise-wide processing networks," *Comput. Chem. Eng.*, vol. 38, pp. 213–223, 2012.

13. C. L. Gargalo, P. Cheali, J. A. Posada, K. V. Gernaey, and G. Sin, "Economic risk assessment of early stage designs for glycerol valorization in biorefinery concepts," *Ind. Eng. Chem. Res.*, vol. 55, no. 24, pp. 6801–6814, 2016.

14. P. E. Murillo-Alvarado, J. M. Ponce-Ortega, M. Serna-Gonzalez, A. J. Castro-Montoya, and M. M. El-Halwagi, "Optimization of pathways for biorefineries involving the selection of feedstocks, products, and processing steps," *Ind. Eng. Chem. Res.*, vol. 52, no. 14, pp. 5177–5190, 2013.

15. Q. Jin, H. Zhang, L. Yan, L. Qu, and H. Huang, "Kinetic characterization for hemicellulose hydrolysis of corn stover in a dilute acid cycle spray flow-through reactor at moderate conditions," *Biomass and Bioenergy*, vol. 35, pp. 4158–4164, Oct. 2011.

16. R. Morales-Rodriguez, K. V. Gernaey, A. S. Meyer, and G. Sin, "A mathematical model for simultaneous saccharification and co-fermentation (SSCF) of C6 and C5 sugars," *Chinese J. Chem. Eng.*, vol. 19, pp. 185–191, Apr. 2011.

17. Y. S. H. Najjar, "Gas turbine cogeneration systems: A review of some novel cycles," *Appl. Therm. Eng.*, vol. 20, pp. 179–197, Feb. 2000.

18. S. Shahhosseini, "Simulation and optimisation of PHB production in fed-batch culture of *Ralstonia eutropha*," *Process Biochem.*, vol. 39, no. 8, pp. 963–969, 2004.

19. N. Leksawasdi, E. L. Joachimsthal, and P. L. Rogers, "Mathematical modelling of ethanol production from glucose/xylose mixtures by recombinant *Zymomonas mobilis*," *Biotechnol. Lett.*, vol. 23, pp. 1087–1093, 2001.

20. E. C. Rivera, A. C. Costa, D. I. P. Atala, F. Maugeri, M. R. W. Maciel, and R. M. Filho, "Evaluation of optimization techniques for parameter estimation: Application to ethanol fermentation considering the effect of temperature," *Process Biochem.*, vol. 41, pp. 1682–1687, Jul. 2006.

21. M. M. El-Halwagi, "Chapter 1: Introduction to sustainability, sustainable design, and process integration," in *Sustainable Design through Process Integration*, pp. 1–14, Eds. Amsterdam, Netherlands: Elsevier, 2012.

22. M. Vázquez-Ojeda, J. G. Segovia-Hernández, and J. M. Ponce-Ortega, "Incorporation of mass and energy integration in the optimal bioethanol separation process," *Chem. Eng. Technol.*, vol. 36, no. 11, pp. 1865–1873, 2013.
23. J. M. Naranjo, C. A. Cardona, and J. C. Higuita, "Use of residual banana for polyhydroxybutyrate (PHB) production: Case of study in an integrated biorefinery," *Waste Manag.*, vol. 34, pp. 2634–2640, 2014.
24. J. Moncada, C. A. Cardona, and L. E. Rincón, "Design and analysis of a second and third generation biorefinery: The case of castorbean and microalgae," *Bioresour. Technol.*, vol. 198, pp. 836–43, Dec. 2015.
25. J. C. Solarte-Toro, Y. Chacón-Perez, and C. A. Cardona Alzate, "Comparison of biomethane and syngas production as energy vectors for heat and power generation from palm residues: A techno-economic, energy and environmental assessment," in *5th International Conference on Sustainable Solid Waste Management*, 2017, p. Session XIX.
26. S. Kumneadklang, S. Larpkiattaworn, and C. Niyasom, "Bioethanol production from oil palm frond by simultaneous saccharification and fermentation," *Energy Procedia*, vol. 79, pp. 784–790, Nov. 2015.
27. C. Bravo-Bravo, J. G. Segovia-Hernández, C. Gutiérrez-Antonio, A. L. Durán, A. Bonilla-Petriciolet, and A. Briones-Ramírez, "Extractive dividing wall column: Design and optimization," *Ind. Eng. Chem. Res.*, vol. 49, no. 8, pp. 3672–3688, 2010.
28. J. A. Dávila, V. Hernández, E. Castro, and C. A. Cardona, "Economic and environmental assessment of syrup production. Colombian case," *Bioresour. Technol.*, vol. 161, pp. 84–90, 2014.
29. A. C. Kokossis and A. Yang, "On the use of systems technologies and a systematic approach for the synthesis and the design of future biorefineries," *Comput. Chem. Eng.*, vol. 34, no. 9, pp. 1397–1405, 2010.
30. C. A. García, R. Betancourt, and C. A. Cardona, "Stand-alone and biorefinery pathways to produce hydrogen through gasification and dark fermentation using *Pinus Patula*," *J. Environ. Manage.*, vol. 203, pp. 1–9, 2015.
31. C. A. García, J. Moncada, V. Aristizábal, and C. A. Cardona, "Techno-economic and energetic assessment of hydrogen production through gasification in the Colombian context: coffee cut-stems case," *Int. J. Hydrogen Energy*, vol. 42, pp. 5849–5864, 2017.
32. C. A. Cardona Alzate, J. C. Solarte-Toro, and Á. G. Peña, "Fermentation, thermochemical and catalytic processes in the transformation of biomass through efficient biorefineries," *Catal. Today*, vol. 302, pp. 61–72, 2018.
33. R. Smith, *Chemical Process Design and Integration*, Eds. McGraw-Hill, 2005.
34. M. O. S. Dias, A. V. Ensinas, S. A. Nebra, R. Maciel Filho, C. E. V. Rossell, and M. R. W. Maciel, "Production of bioethanol and other bio-based materials from sugarcane bagasse: Integration to conventional bioethanol production process," *Chem. Eng. Res. Des.*, vol. 87, no. 9, pp. 1206–1216, 2009.
35. J. Nagel, "Determination of an economic energy supply structure based on biomass using a mixed-integer linear optimization model," *Ecol. Eng.*, vol. 16, pp. 91–102, 2000.
36. J. A. Quintero, J. Moncada, and C. A. Cardona, "Techno-economic analysis of bioethanol production from lignocellulosic residues in Colombia: A process simulation approach," *Bioresour. Technol.*, vol. 139, pp. 300–307, Jul. 2013.

37. P. Lv, Z. Yuan, L. Ma, C. Wu, Y. Chen, and J. Zhu, "Hydrogen-rich gas production from biomass air and oxygen/steam gasification in a downdraft gasifier," *Renew. Energy*, vol. 32, no. 13, pp. 2173–2185, 2007.
38. A. M. Y. Razak, *Industrial Gas Turbines: Performance and Operability*. Elsevier, 2007, pp. 1–624.
39. L. F. Gutiérrez, Ó. J. Sánchez, and C. A. Cardona Alzate, "Analysis and design of extractive fermentation processes using a novel short-cut method," *Ind. Eng. Chem. Res.*, vol. 52, no. 36, pp. 12915–12926, 2013.
40. K. L. Wasewar, A. A. Yawalkar, J. A. Moulijn, and V. G. Pangarkar, "Fermentation of glucose to lactic acid coupled with reactive extraction: A review," *Ind. Eng. Chem. Res.*, vol. 43, no. 19, pp. 5969–5982, 2004.
41. N. I. Canabarro, C. Alessio, E. L. Foletto, R. C. Kuhn, W. L. Priamo, and M. A. Mazutti, "Ethanol production by solid-state saccharification and fermentation in a packed-bed bioreactor," *Renew. Energy*, vol. 102, pp. 9–14, 2017.
42. Y. A. Pisarenko, L. A. Serafimov, C. A. Cardona, D. L. Efremov, and A. S. Shuwalov, "Reactive distillation design: Analysis of the process statics," *Rev. Chem. Eng.*, vol. 17, no. 4, pp. 253–327, 2001.
43. L. A. Serafimov, Y. A. Pisarenko, and N. N. Kulov, "Coupling chemical reaction with distillation: Thermodynamic analysis and practical applications," *Chem. Eng. Sci.*, vol. 54, no. 10, pp. 1383–1388, 1999.
44. M. S. Peters, K. D. Timmerhaus, and R. E. West, "Chapter 14: Materials Transfer, Handling, and Treatment Equipment-Design and Costs," in *Plant Design and Economics for Chemical Enginners*, pp. 478–570. McGraw-Hill, 2004.
45. G. D. Ulrich and P. T. Vasudevan, "How to estimate utility costs," *Chem. Eng.*, vol. 113, no. 4, pp. 66–69, 2006.
46. J. Moncada, J. A. Tamayo, and C. A. Cardona, "Integrating first, second, and third generation biorefineries: Incorporating microalgae into the sugarcane biorefinery," *Chem. Eng. Sci.*, vol. 118, pp. 126–140, 2014.
47. V. Hernández, J. M. Romero-García, J. A. Dávila, E. Castro, and C. A. Cardona, "Techno-economic and environmental assessment of an olive stone based biorefinery," *Resour. Conserv. Recycl.*, vol. 92, pp. 145–150, 2014.
48. J. Moncada, C. A. Cardona, J. C. Higuita, J. J. Vélez, and F. E. López-Suarez, "Wood residue (Pinus patula bark) as an alternative feedstock for producing ethanol and furfural in Colombia: experimental, techno-economic and environmental assessments," *Chem. Eng. Sci.*, vol. 140, pp. 309–318, Feb. 2016.

5

Environmental Assessment of Biorefinery Systems

According to the European Commission, environmental assessment is defined as the methodology guaranteeing that the environmental consequences of decisions are considered before these decisions are made [1]. An environmental assessment can be applied to any project, a motorway, a building, a factory, a process, etc. The purpose is to identify the environmental effects, to propose solutions in order to mitigate problems, and to predict the possible effects. The environmental assessment is a methodological tool to identify, assess, and make decisions with respect to the environmental impacts generated in a process or biorefinery. An environmental assessment applied to a process or biorefinery should consider the important aspects such as flows of feedstock, reagents and products, operating conditions, fuels, and land use, among others, in order to define the limits and goals of environmental analysis and to know the possible bottlenecks of the system. The sustainability of a process or biorefinery depends on four essential pillars: technical, economic, environmental, and social analyses. In this sense, this chapter presents a comprehensive overview of methodological approaches that are applied to this type of systems to assess the environmental impact of a product. In the Sections 5.1–5.4, the methodologies that are more commonly used and some examples of applicability are described.

5.1 Life Cycle Assessment

Life cycle assessment (LCA) is defined as a method to assess and determine the environmental impact of goods and products throughout its full life cycle, from extraction until recycling or disposal as waste. It is the methodological approach extended worldwide to carry out global assessments (economic, energy, environmental, and social analysis) [2]. LCA is used to calculate the environmental burdens of complex and multidisciplinary processes. This method can be applied to sectors such as building, agriculture, transport, and engineering, among others. ISO 14040:2006, ISO 14044:2006, International Reference Life Cycle Data System, ISO 14025:2006, and so on are the international standards and guidelines that indicate how to carry out an LCA from the beginning to the end, applied to any product or system. In addition, software and databases have been developed to carry out the LCA. The most important are GaBi, EcoManager, LCA1, LCAD, LIMS, LMS Eco-Inv. Tool, Bio-Grace, SimaPro, TEMIS, and Umcon, among others.

Uihlein and Schebek (2009) [3], Cherubini and Jungmeier (2010) [4], Cavalett et al. (2012) [5], De Meester (2013)[6], Karka et al. (2014) [7], and WBCSD Chemicals [8] have presented the relevance of this methodology applied in the assessment of processes and biorefineries. The assessment of this type of systems using LCA methodology considers four important steps that are described in Sections 5.1.1–5.1.4.

5.1.1 Goal and Scope Definition

This stage answers to the following questions: What is the goal to achieve? What processes do the LCA include and how do they work? [9].

This step determines the methodological options that an LCA should consider. A clear goal definition is necessary for correct interpretation of results. Based on ISO 14040 and ISO 14044, the goal is composed of (i) reasons for the study (formulation of research questions), (ii) accurate definition of the product, (iii) functional unit definition, (iv) description of the system boundaries, (v) requirements and assumptions, and (vi) expected public. The formulation of the questions is an important part in goal definition. For the biorefineries case, these can be formulated considering the feedstock use and number of products. In addition, in the goal and scope definition, the time horizons of the LCA should be specified.

Figure 5.1 shows an example of the boundaries that can be applied to a system according to the goal in its LCA. The first boundary called "gate to gate" involves only the product(s) processing obviating the environmental load of feedstock(s) and the product use. For the case where the LCA considers the environmental impact of feedstock(s), the boundary is known as "cradle to gate." On the other hand, the third boundary is "cradle to grave', which makes the analysis from the origin of feedstock(s) followed by the processing, until the treatment of product(s) as waste (end of life). Finally, the last boundary is called "cradle to cradle" where the LCA considers that the product(s) is recycled and is fed again to the system as feedstock(s).

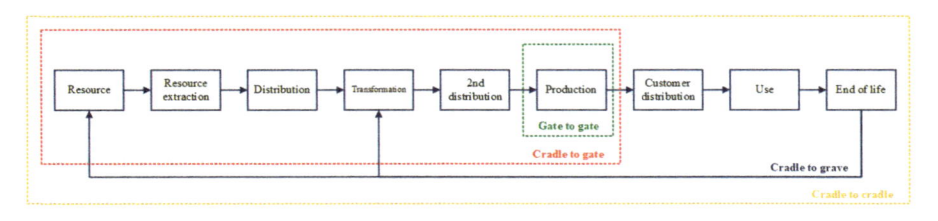

FIGURE 5.1
Example of system boundaries in LCA.

After the system boundaries are chosen, another important aspect in the goal and scope definition of LCA is the selection of functional unit. The functional unit is defined as the measure to quantify the environmental impact of goods and products, and it can be referred to as an amount of mass or energy [9]. It allows having a reference point in inputs and outputs of the system; additionally, standards and guidelines define the selection of functional unit as a key aspect in the comparison of study cases and result interpretation.

Given that the biorefinery considers more than one product, the selection of the functional unit is a complex task. The functional unit is rigorously related with the goal definition; therefore, depending on the research questions, the functional unit will be selected. For example, when the goal is the comparison between the products of a biorefinery, the products or the function of the products can be the functional unit. In a biorefinery, the use of more than two functions is normal and can be used to compare stand-alone processes and biorefineries. For study cases where various goals are considered, different functional units should be used in order to fulfill the goals.

Once the goal and scope definition are made, the next step is to collect data about the system, namely, to do a life cycle inventory (LCI) assessment.

5.1.2 Life Cycle Inventory

LCI is the step where the determination of input and output flows and the calculation of emissions for different environmental resources are made (energy use, resource extraction, or emission to an environmental behavior) [5]. Each process that considers a product should have a data set or compilation related to the function or product generated [9]. In LCA, the data compilation and collection are the most challenging, and work- and time-consuming steps, but it is the key to obtain useful data [9,10]. Two types of data are considered in LCI. The first type is background data, which are frequently used for common processes in almost all studies, for example, transport, waste management, energy supply, and commodity chemicals and materials [9]. The other data type is foreground data, which refer to those that need to be collected or acquired from databases for specific and specialized production systems.

5.1.3 Life Cycle Impact Assessment

The goal of life cycle impact assessment (LCIA) is to classify and evaluate the results and to compare with standards, models, or known tools. The LCIA associates the inventory data with specific environmental impacts and with indicators of these categories in order to understand and assess the magnitude of the potential environmental impacts (PEIs) for any study case [11]. In this step, an iterative process can be carried out because the results from previous steps are assessed according to the goal and scope to obtain recommendations and conclusions and determine if the goal was reached or it should be modified [11,12].

The LCIA must consider operating and mandatory elements [13]. The operating elements allow obtaining the LCIA profile: the selection of impact categories, category indicators and characterization models, the allocation results of LCIA (classification), and the calculation results of category indicator (characterization), among others. The operating elements are normalization, weighting, and grouping [11,14].

5.1.4 Interpretation of Results

The interpretation of results is the last step in an LCA, in which the results obtained in LCI and LCIA are grouped and analyzed in order to generate decision elements [15]. The findings of interpretation can offer conclusions and recommendations on decision-making, coherent with the goal and scope of the study. This step aims to offer an accessible, complete and coherent reading of the results presentation of an LCA in order to identify the vulnerable stages of process with a considerable environmental impact. [11,14].

The LCA methodology requires the allocation or multifunctionality concept when a multi-output system is considered (ISO 14040:2006; ISO 14044:2006) [5] and a biorefinery is a typical case due to the multiple inputs (feedstocks and reagents) and outputs. Here, the selection of the method to solve the multifunctionality of the system plays an important role in the results [16]. The multifunctionality or allocation procedure considers two ways [12]. The first one is by expanding the system to each product, where the environmental load is calculated with respect to each product. However, this alternative is complex and laborious. The second one consists in the partitioning of inputs and the environmental impacts of the system to the main products taking into account weight factors or allocated burdens [7,12]. This way gives weight to the environmental impact to each product with respect to its nature (i.e., main product, co-product, by-product, waste), its economic cost, or in other cases physical characteristics of the feedstock. When an allocation is made with respect to a given factor, it should be justified with solid arguments and presented in detail in the final report [12]. In this sense, the allocation topic is very relevant due to the possibility of several scenarios to assess the different impacts of the products [7].

5.2 Potential Environmental Impact

The PEI allows determining the effect that would have a quantity of mass and energy if these were emitted in the environment [17–19]. It can be calculated by quantifying the effect of a release, from a process or biorefinery, that has not happened or is going to happen [18]. It cannot be measured directly with an instrument because it is a conceptual quantity, but there are measurable quantities to be estimated [20].

The PEI calculation is proposed as a global mass and energy balance where environmental effects on the materials that enter and leave the process or biorefinery and the energy consumption are considered. Figure 5.2 shows the boundaries of the system around the production process. The expression of PEI balance is given by the following equation:

$$\frac{\partial I_{syst}}{\partial t} = I_{in}^{(p/b)} + I_{in}^{(eg)} - I_{out}^{\left(\frac{p}{b}\right)} - I_{out}^{(eg)} - I_{we}^{\left(\frac{p}{b}\right)} - I_{we}^{(eg)} + I_{gen}^{(t)} \tag{5.1}$$

where:

I_{syst} is the PEI inside the system (process or biorefinery as well as energy generation).

$I_{in}^{(p/b)}$ and $I_{out}^{(p/b)}$ are the PEIs entering and leaving the process or biorefinery (PEI/h).

$I_{in}^{(eg)}$ and $I_{out}^{(eg)}$ are the PEIs entering and leaving the energy generation (PEI/h).

$I_{we}^{(p/b)}$ and $I_{we}^{(eg)}$ are the PEIs emitted from the process or biorefinery and energy generation associated with waste energy (PEI/h).

$I_{gen}^{(t)}$ is the total PEI generated inside the system normalized to the production rate (PEI/kg products).

The term on the left side of equation is zero for steady-state systems. The outputs of PEI associated with waste energy lost from the biorefinery and energy generation can be considered negligible due to the low impacts in comparison with other streams. Additionally, the input rates of energy

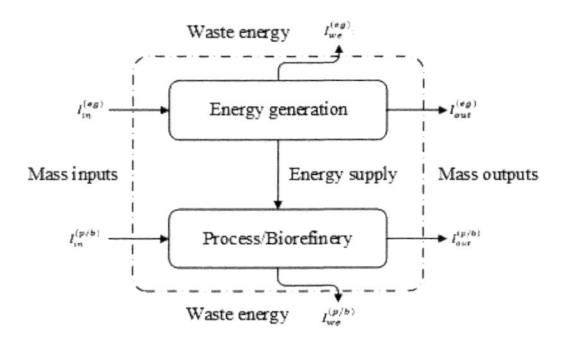

FIGURE 5.2
Boundaries of system around the production process to determine PEI. (Adapted from [18].)

generation are also insignificant for reasons exposed by Young and Cabezas [20]. In this sense, the real expression of PEI balance is given as follows:

$$I_{in}^{(p/b)} - I_{out}^{\left(\frac{p}{b}\right)} - I_{out}^{(eg)} + I_{gen}^{(t)} = 0 \tag{5.2}$$

In general terms, the PEI balance is defined as follows:

$$I_{in}^{(t)} - I_{out}^{(t)} + I_{gen}^{(t)} = 0 \tag{5.3}$$

where:

$I_{in}^{(t)}$ is the total PEI entering the system (PEI/h).

$I_{out}^{(t)}$ is the total PEI leaving the systems (PEI/h).

The input PEI index can be determined by known quantities as follows:

$$I_{in}^{(t)} = \sum_i^{EnvCat} \alpha_i I_{i,in}^{(t)} = \sum_i^{EnvCat} \alpha_i \sum_j^{Streams} M_{j,in} \sum_k^{Comps} x_{kj} \psi_{ki}^s \tag{5.4}$$

where:

α_i is the weighting factor of impact category i.

$I_{i,in}^{(t)}$ is the total PEI entering the system of impact category i (PEI/h).

$M_{j,in}$ is the mass flow of stream j into the system.

x_{kj} is the mass fraction of chemical k in stream j.

The output PEI index can be determined in a similar way to the input PEI index as indicated in Equation 5.5. While estimating the output PEI index, only non-product streams are taken into account [18].

$$I_{out}^{(t)} = \sum_i^{EnvCat} \alpha_i I_{i,out}^{(t)} = \sum_i^{EnvCat} \alpha_i \sum_j^{NPStreams} M_{j,out} \sum_k^{Comps} x_{kj} \psi_{ki}^s \tag{5.5}$$

where:

$I_{i,out}^{(t)}$ is the total PEI leaving the system of impact category i (PEI/h).

$M_{j,out}$ is the mass flow of stream j out of the system.

From Equation 5.3, some indexes can be calculated considering mass and energy inputs and output streams from the process or biorefinery and the global environmental impact of chemical products used as reagents or released as pollutants. For assessing the environmental performance of a process or biorefinery, two types of indexes are used: PEI output indexes and PEI generation indexes, which are generally represented in PEI/h.

Nevertheless, these indexes can be transformed in PEI/kg of product taking into account the mass flow of product(s). The following equations show this transformation:

$$\widehat{I_{out}^{(t)}} = \frac{I_{out}^{(t)}}{\displaystyle\sum_p^{ProdStreams} P_p} \tag{5.6}$$

$$\widehat{I_{gen}^{(t)}} = \frac{I_{gen}^{(t)}}{\displaystyle\sum_p^{ProdStreams} P_p} \tag{5.7}$$

The PEI of a process or biorefinery can be assessed through the tool and methodology called waste reduction (WAR) algorithm [18–20], which has been driven by the United States Environmental Protection Agency (US EPA). WAR is not a LCA because it considers only the LCA of the synthesizing stage of product(s), leaving aside the important aspects, before and after the process or biorefinery. For example, a complete LCA takes into account aspects such as feedstock availability, product(s) demand, use, and recycle. However, the WAR methodology is useful to do a pre-feasibility study in the design stage of a process or biorefinery, or in the reconsideration of a current process or biorefinery where this type of analysis could be considered an essential part of the design to select alternatives [18]. The determination of PEI using WAR algorithm is mainly applicable to the comparison of processes or biorefineries (including the scenario) because there are not reported numerical ranges that provide an idea about its good or bad performance. Systems with lower PEI are more environmentally friendly [18]. Comparison can be made considering different scenario aspects such as type and distribution of feedstocks, technologies, and configurations, among others.

The WAR algorithm considers eight environmental impact categories: human toxicity potential by ingestion, human toxicity potential by exposure, aquatic toxicity potential, terrestrial toxicity potential, global warming potential (GWP), ozone depletion potential, photochemical oxidation potential, and acidification potential [18–20]. Table 5.1 shows some examples of the calculation of PEI applied to biorefineries.

5.3 Water Footprint

The water footprint (WF) is defined as an indicator that allows quantifying the consumed water in the production of goods and services [26,27]. The WF is divided into three components: green, blue, and gray waters, which complete the assessment of Water Footprint Network (WFN) according

TABLE 5.1

Examples of Biorefineries Assessed from an Environmental Perspective Using PEI

Biorefinery	Scenarios	Environmental Results	Reference
Microalgae and sugarcane biorefinery	Scenario 1: Sugar, ethanol, and electricity. Scenario 2: Extension of scenario 1 that also considers *Chlorella* sp. cultivation using CO_2-rich streams derived from fermentation and cogeneration systems. After microalgae cultivation and harvesting, the extracted oil is used as the raw material to obtain biodiesel and glycerol as additional products.	Scenario 2 has the best performance from environmental and economic perspectives. The PEI is 44.99 PEI/ton of products, with the potential to reduce CO_2 emissions by 39% compared with scenario 1.	[21]
Castorbean and microalgae biorefinery	Products: Polyol, ethylene glycol, omega-3 acid, biodiesel, methanol, and heat and power. Three scenarios based on different levels of mass and energy integration: Scenario 1: Products and internal mass integration of process streams. Scenario 2: Products and internal mass integration, as well as external mass integration (water sources) and heat integration (pinch). Scenario 3: Products and mass and heat integration as well as energy supply by the cogeneration system.	The PEIs for scenarios 1, 2, and 3 are 140, 48, and 35 PEI/ton of products, respectively. The environmental impacts of scenario 2 are 2.9 times reduced, whereas the impacts of scenario 3 are 4 times reduced compared with scenario 1.	[22]
Rejected banana biorefinery	Scenario 1: Polyhydroxybutyrate (PHB). Scenario 2: Glucose, ethanol, and PHB. Scenario 3: Scenario 2 plus mass and energy integration.	kg total PEI/kg PHB and kg CO_2/kg PHB Scenario 1: 0.5 and 2.5 Scenario 2: 0.35 and 2.2 Scenario 3: 0.25 and 1.5	[23]
Theobroma grandiflorum (copoazu) biorefinery	Products: pasteurized pulp, antioxidant extract, biofertilizer, biogas, oil seed, essential oil, ethanol, and PHB. Scenario 1: Base case. Scenario 2: Base case and energy integration. Scenario 3: Base case and energy integration and cogeneration.	The PEIs for scenario 1, 2 and 3 are 0.15, 0.14, and 0.001 PEI leaving the system/kg product.	[24]
Andes berry (*Rubus glaucus* Benth.) biorefinery	Products: Phenolic compound extracts, ethanol and xylitol Scenario 1: Base case. Scenario 2: Base case and mass integration Scenario 3: Base case and mass and energy integration Scenario 4: Base case and cogeneration system	PEI/kg of product for leaving PEI and PEI/kg of product for generated PEI Scenario 1: 6.49 and −33.8 Scenario 2: 6.18 and −34.1 Scenario 3: 0.603 and −6.33 Scenario 4: 0.557 and −6.37	[25]

to ISO Standard 14046 [26]. The relevance of WF is focused on the goal of reducing the water consumption to the strictly necessary in order to have more environmentally friendly processes, operations, and markets. Its importance is linked to the majority of producers and consumers who do not know the real freshwater consumption for obtaining products and services [27].

The functional unit of WF should be defined according to the scope of the system. It is normally expressed as the volume of water used to obtain one unit of product (m^3/ton or kg) or the volume of water per year referred to an area (e.g., country, town, etc.), community, or individual (m^3/year) [26,27].

The WF presents a framework of application with two different approaches [26,28]:

i. The WFN, which is the original version and considers steps such as scope and goal, accounting, sustainability evaluation, and answer formulation in order to quantify the water content in a determined system

ii. The LCA approach, which is carried out through steps such as goal and scope definition, LCI, LCIA, and interpretation of results (discussed earlier).

Both approaches are valid to quantify the consumed freshwater in a system through WF. However, the WF is not a single value three types of WF are defined by colors (green, blue and grey).

The use of WF in the biorefinery context allows determining the water needs from the feedstock obtaining until the final disposal of wastewater streams of all processes considered in the system. Table 5.2 shows some examples of the calculation of WF applied to biorefineries. The WF is a useful tool to do a distinction between the types of water used in the production process. The water can be domestic, industrial, and of agricultural type. In this sense, blue water presents a considerable contribution when industrial and domestic sectors are treated, whereas for agricultural sector, green water is the most representative. The three components are quantified in a separate way. The WF of a product or service is the sum of three water volumes (m^3), as shown in Equation 5.8. Each component fulfills a mission in the WF because each one is associated with different economic, environmental, and social impacts.

$$WF = WF_{green} + WF_{blue} + WF_{gray} \tag{5.8}$$

Each footprint plays a different role and its contribution is related to the studied system. The WF_{green} is relevant when an agricultural or forest product is obtained because the water incorporated in harvested culture, evaporated from soil, absorbed by roots of plants and transpired by the

TABLE 5.2

Examples of Biorefineries Assessed from an Environmental Perspective Using WF

Biorefinery	Scenarios	Environmental Results	Reference
Rejected banana biorefinery	Scenario 1: Polyhydroxybutyrate (PHB). Scenario 2: Glucose, ethanol, and PHB. Scenario 3: Scenario 2 plus mass and energy integration.	m^3 H_2O/ton PHB Scenario 1: 12,500 Scenario 2: 11,500 Scenario 3: 7,500	[23]
Algal biorefinery	1,000 kg of algal biomass Plant A: 247.2 L biodiesel, 21.7 kg glycerin and 700 kg algal meal. Plant B: 197.8 L biodiesel, 17.4 kg glycerin, 43.5 kg omega-3 fatty acids, and 700 kg algal meal.	WF to produce 10–13 million gallons of algal biodiesel per annum from plants A and B. 39.9 Annual water use biomass production (billion liters) Annual water use for processing by the facility (billion liters) Plant A: 0.0094 Plant B: 0.0076	[33]
Corn stover and perennial grass (switchgrass and *Miscanthus*) biorefinery	Renewable diesel blendstock (RDB). The biorefinery process includes a blue water and gray WF	Water consumption (gallon/gallon RDB) Cooling 12.23, aeration 0.03, bioreactor vent 0.87, pretreatment 0.16, enzymatic hydrolysis 0.18, boiler blowdown vent 0.17, upgrading flue gas 0.01, upgrading produced water 0.06, wastewater treatment (WWT) evaporation 0.19, WWT brine 0.25, combustor stack 4.81, total 18.90	[34]
Sugarcane biorefinery	Scenario 50/50: Anhydrous ethanol, sugar, and power. Sugarcane juice split of 50:50 for ethanol and sugar production, respectively. Scenario 75/25: Scenario 50/50; however, juice split of 75:25 (ethanol:sugar) RS-C: Butanol integrated to anhydrous ethanol, sugar, and power. Sugarcane juice split in a proportion of 50:25:25 for the production of ethanol, sugar, and acetone, butanol, and ethanol (ABE), respectively. A regular *Clostridium* strain is considered for butanol production. Butanol is commercialized as a chemical.	Water consumption for dilution of sugarcane juice (m^3/h) Scenario 50/50: N.R. Scenario 75/25: N.R. Scenarios RS-C and RS-B: 198.1 Scenarios MS-C and MS-B: 131.2 Corresponds to the amount of water to be combined with the water stream available in the distillation unit of the butanol plant (18.2 and 22.6 m^3/h).	[35]

(Continued)

TABLE 5.2 (*Continued*)

Examples of Biorefineries Assessed from an Environmental Perspective Using WF

Biorefinery	Scenarios	Environmental Results	Reference
	MS-C: 50:25:25 (ethanol:sugar:ABE). Mutant *Clostridium* strain with improved butanol yield. Butanol is commercialized as a chemical.		
	RS-B: 50:25:25 (ethanol:sugar:ABE). Regular *Clostridium* strain. Butanol is commercialized as an automotive fuel.		
	MS-B: 50:25:25 (ethanol:sugar:ABE). Mutant Clostridium strain with. Butanol is commercialized as an automotive fuel.		

N.R. non-reported.

crop must be taken into account [27,29]. WF_{green} is considered only when the analysis considers agricultural processes, and it is defined by the following equations [30]:

$$\text{If } ET \geq P_{eff}, \; WF_{green} = P_{eff} \tag{5.9}$$

$$\text{If } ET < P_{eff}, \; WF_{green} = ET \tag{5.10}$$

where:

ET is the evapotranspiration occurred during crop growth (P).

P_{eff} is the effective rainfall occurred during crop growth (P).

WF_{blue} is the water taken from superficial or underground water resources, which is incorporated into soil by crop irrigation. This type of water is expensive because it is used in other activities to satisfy the human needs, generating competition. Additionally, it includes other costs such as pumping, machines for management and distribution, and energy requirements.

The requirements of blue water depend on crop characteristics, if is tolerant to water deficit, the irrigation technique and green water [26]. Blue water is used if green water is insufficient. Blue water is calculated as follows [30]:

$$\text{If } ET \geq P_{eff}, \; WF_{blue} = ET - P_{eff} \tag{5.11}$$

$$\text{If } ET < P_{eff}, \; WF_{blue} = 0 \tag{5.12}$$

WF_{grey} is defined as the volume of freshwater that is required for assimilating pollutants and fulfilled with water quality standards of streams spilled to the atmosphere in a production process [27]. Gray water is not a real volume used in the production process, but it is the volume of water needed to reestablish the water quality after it is polluted [26]. The determination of gray water is carried out through Equations 5.13 and 5.14 [30]. WF_{gray} is mainly composed of fertilizers used and their nutrient load, industrial wastewaters, and effluents [31,32]. In agriculture, nitrogen fertilizers are the most commonly used in soil, and their dosage depends on the crop type and local allowed standards [26].

$$WF_{gray} = \frac{L_{add}}{C_{max} - C_{nat}} \tag{5.13}$$

where:

WF_{gray} $(m^3/year)$ is the gray water footprint.

L_{add} (Gg/year) is the pollutant load.

C_{max} (mg/dm^3) is the maximum concentration (acceptable) of the pollutant.

C_{nat} (mg/dm^3) is the natural concentration of the pollutant in the water body (receiver).

$$L_{add} = L - C_{nat} * Q_{act} \tag{5.14}$$

where:

L (Gg/year) is the total pollutant load.

Q_{act} (m^3/yr) is the actual basin discharge.

5.4 Greenhouse Gases

The goal of greenhouse gases (GHGs) is to calculate the GHG emissions of a supply chain. This methodology can be used to calculate the GHG emissions of the production of feedstock, transformation processes, and distribution of the final product of a process or biorefinery. Due to the fact that GHG emissions can be calculated in all stages of a supply chain, the methodology suggests to delimit the system [36].

The GHGs, according to US EPA, are defined as gases that catch heat in the atmosphere promoting the well-known global warming [37]. The six categories of GHGs identified using Kyoto Protocol are carbon dioxide (CO_2), methane (CH_4), nitrous oxide (N_2O), fluorinated gases (i.e., hydroflu-orocarbons, perfluorocarbons, and sulfur hexafluoride (SF_6)). These gases are released to the atmosphere through industrial and agricultural activities, for example, CO_2 enters to the atmosphere through activities such as burning of waste, wood, and fossil fuels, and as products of some chemical reactions. CH_4 is generated during some agricultural activities and fossil fuel transport and production. N_2O is produced during combustion of waste and fossil fuels. Finally, fluorinated gases are produced from industrial processes [37]. In multiportfolio processes such as biorefineries, the activities mentioned previously can occur. For this reason, the calculation of GHGs is relevant to determine the environmental performance of this type of systems. Table 5.3 shows some examples of the calculation of GWP applied to biorefineries.

The methodology for calculating the GHGs considers some steps: (i) to identify the emission sources, (ii) to collect the activity data, (iii) to define the units of measure the data, and (iv) to determine the effect of each GHG to the global warming through the calculation of GWP.

The activity data should be translated into one or more quantities of GHGs. A single activity can produce more than one type of GHGs, for example, the natural gas burning produces CO_2, CH_4, and N_2O. This translation is carried out using one or more emissions factors (EFs). The EF is defined as the ratio of the amount of the emitted GHGs to a given unit of activity. The units of EF are a unit of the amount of GHGs over a functional unit of activity (e.g., g CO_2 per kWh, kg CH_4 per kg or ton, g N_2O per liter, etc.) The EF are provided by government institutions as US EPA, and these can be obtained from academic and scientific papers or official reports.

A GWP is the ratio of the effect of an amount of GHGs on climate change to an equal amount of CO_2, as indicated in Equation 5.15 [38]. It is generally represented by a period of 100 years; it is expressed in kg-ton-lb CO_2-eq and is continually refined. Finally, the CO_2 has a GWP of 1.

$$GWP = \frac{\sum_{i=1}^{I} m_i^{out} * EF_{CO_2, i}}{m_{prod}} \tag{5.15}$$

where:

m_i^{out} (kg) is the output mass of the component i.

$EF_{CO_2, i}$ $(kg\ CO_2/kg\ i)$ is the emission factor.

M_{prod} (kg) is the product mass.

TABLE 5.3

Examples of Biorefineries Assessed from an Environmental Perspective Using GWP

Biorefinery	Scenarios	Environmental Results	Reference
Sugarcane biorefinery	Scenarios proposed as a function of feedstock distribution and Technologies. Scenario 1: Colombian base case, sugar, fuel ethanol, and electricity. Scenarios 2 and 3: Sugar, fuel ethanol, electricity, PHB, and anthocyanins.	The GHG emissions represented as kg CO_2 eq per kg of processed cane are 0.52, 1.30 and 0.78 for scenarios 1, 2, and 3, respectively.	[39]
Corn stover biorefinery	Scenario 1: Dilute acid followed by overliming. Scenario 2: Dilute acid followed by ammonia addition Scenario 3: Two-stage pretreatment followed by ammonia addition Scenario 4: Dilute acid followed by membrane separations	Scenario 1: 22 g CO_2 eq per kg-DB (dry biomass). Scenario 2: 18.9 g CO_2 eq per kg-DB. Scenario 3: 0.76 g CO_2 eq per kg-DB. Scenario 4: 6.2 g CO_2 eq per kg-DB.	[40]
Switchgrass biorefinery	Scenario 1: Ethanol, phenols, methane, electricity, and heat. Scenario 2: Fossil reference system producing the same products/services from fossil sources	Scenario 1: 65 kt CO_2-eq/year. Scenario 2: 281 kt CO_2-eq/year.	[41]
Multi-feedstock biorefinery	Assessment of PEIs of producing maize, grass-clover, ryegrass, and straw from winter wheat as biomass feedstocks for biorefinery	Net GWP_{100} (including soil C change). Maize: 3119 kg CO_2 eq/ha, 315 kg CO_2 eq/t DM (dry matter). Grass-clover: 2728 kg CO_2 eq/ha, 354 kg CO_2 eq/t DM. Ryegrass: 3588 kg CO_2 eq/ha, 410 kg CO_2 eq/t DM. Winter wheat straw: 492 kg CO_2 eq/ha, 152 kg CO_2 eq/t DM.	[42]
Chlorella vulgaris biorefinery	Scenario 0: Biodiesel (1 t), glycerin (0.05 t) and cake (1.35 t). Scenario 1: Biodiesel (1 t), glycerin (0.05 t), and pyrolysis oil (0.8 t). Scenario 2: Biodiesel (1 t), glycerin (0.05 t), and biogas with 70% CH_4 and 30% CO_2 (0.01 t). Scenario 3: Biodiesel (1 t), glycerin (0.05 t), and pyrolysis oil (0.8 t). Scenario 4: Biodiesel (1 t), glycerin (0.05 t), and biogas with 70% CH_4 and 30% CO_2 (0.01 t)	Emissions by mass (t CO_2 eq/t of biodiesel) and emissions by energy (t CO_2 eq/GJ bioenergy). Scenario 0: 5.12 and 0.08. Scenario 1: 5.10 and 0.10. Scenario 2: 4.88 and 0.12. Scenario 3: 2.38 and 0.05. Scenario 4: 1.85 and 0.05.	[43]

5.5 Conclusions

The selected scheme for a promising biorefinery must include economic and environmental assessments. Initially, in the industry (based not only on biomass but also on oil), the economic factors were the most important. However, environmental factors have played a crucial role in recognition and choice of better process schemes. From the environmental perspective, the biorefineries have been assessed through different methodologies such as PEI, WF, and GHGs. However, industry and academy have proposed the LCA methodology, which is a procedure that groups the methods mentioned earlier in an effective way and considers additional indicators in order to provide a complete and strong environmental analysis. The LCA approach has not been totally developed for its application to biorefineries, which are a complex system. The literature practically does not show an integral LCA analysis for biorefineries, excluding stand-alone processes (close to type I according to the International Energy Agency) that are commonly confused with complex multiproduct biorefineries. This is the main constraint to be solved in the future through more publications in this complex topic to disseminate more and more this strategy.

References

1. European Commission, "Environment," 2017. [Online]. Available: http://ec.europa.eu/environment/eia/index_en.htm. Accessed September 2017.
2. M. Valencia Botero, C. A. Cardona Alzate, and C. Younes Velosa, Cálculo de gases de efecto invernadero en procesos químicos y biotecnológicos con énfasis en cadenas de suministro. 2013.
3. A. Uihlein and L. Schebek, "Environmental impacts of a lignocellulose feedstock biorefinery system: An assessment," *Biomass and Bioenergy*, vol. 33, pp. 793–802, 2009.
4. F. Cherubini and G. Jungmeier, "LCA of a biorefinery concept producing bioethanol, bioenergy, and chemicals from switchgrass," *Int. J. Life Cycle Assess.*, vol. 15, pp. 53–66, 2010.
5. O. Cavalett et al., "Environmental and economic assessment of sugarcane first generation biorefineries in Brazil," *Clean Technol. Environ. Policy*, vol. 14, pp. 399–410, 2012.
6. S. De Meester, "Life cycle assessment in biorefineries: Case studies and methodological development," PhD Thesis. Ghent University, Ghent, Belgium, 2013.
7. P. Karka, S. Papadokonstantakis, K. Hungerbühler, and A. Kokossis, "Environmental impact assessment of biorefinery products using life cycle analysis," *Comput. Aided Chem. Eng.*, vol. 34, pp. 543–548, 2014.

8. WBCSD Chemicals, "Life cycle metrics for chemical products," 2014. Available (online): https://www.wbcsd.org/Projects/Chemicals/Resources/Life-Cycle-Metrics-for-Chemical-Products. Accessed: August 2017.

9. G. Rebitzer et al., "Life cycle assessment. Part 1: Framework, goal and scope definition, inventory analysis, and applications," *Environ. Int.*, vol. 30, no. 5, pp. 701–720, 2004.

10. I. S. De Meester, "Life cycle assessment in biorefineries: Case studies and methodological development," PhD Thesis. Ghent University, Department of Sustainable Organic Chemistry and Technology, 2013.

11. Norma Técnica Colombiana—ISO 14040 (Icontec). *Gestion Ambiental: Analisis de ciclo de vida. Principios y marco de referencia.* Colombia: Instituto Colombiano de Normas Técnicas y Certificación 2007, pp.1–24.

12. G. Finnveden et al., "Recent developments in life cycle assessment," *J. Environ. Manage.*, vol. 91, pp. 1–21, 2009.

13. D. W. Pennington et al., "Life cycle assessment. Part 2: Current impact assessment practice," *Environ. Int.*, vol. 30, no. 5, pp. 721–739, 2004.

14. Norma Técnica Colombiana—ISO 14044 (Icontec). *Gestion Ambiental. Analisis de ciclo de vida. Requisitos y directrices. Requisitos del ciclo de vida.* 2007, pp. 1–50.

15. V. Castellani, S. Sala, and L. Benini, "Hotspots analysis and critical interpretation of food life cycle assessment studies for selecting eco-innovation options and for policy support," *J. Clean. Prod.*, vol. 140, pp. 556–568, 2017.

16. S. Ahlgren et al., "Review of methodological choices in LCA of biorefinery systems—key issues and recommendations," *Biofuels, Bioprod. Biorefining*, vol. 6, no. 3, pp. 606–619, 2012.

17. Q. Chen and X. Feng, "Potential environmental impact (PEI) analysis of reaction processes," *Comput. Aided Chem. Eng.*, vol. 15, pp. 748–753, 2003.

18. D. Young, R. Scharp, and H. Cabezas, "The waste reduction (WAR) algorithm: Environmental impacts, energy consumption, and engineering economics," *Waste Manag.*, vol. 20, pp. 605–615, 2000.

19. H. Cabezas, J. C. Bare, and S. K. Mallick, "Pollution prevention with chemical process simulators: The generalized waste reduction (WAR) algorithm," *Comput. Chem. Eng.*, vol. 21, pp. S305–S310, 1997.

20. D. M. Young and H. Cabezas, "Designing sustainable processes with simulation: The waste reduction (WAR) algorithm," *Comput. Chem. Eng.*, vol. 23, pp. 1477–1491, 1999.

21. J. Moncada, J. A. Tamayo, and C. A. Cardona, "Integrating first, second, and third generation biorefineries: Incorporating microalgae into the sugarcane biorefinery," *Chem. Eng. Sci.*, vol. 118, pp. 126–140, 2014.

22. J. Moncada, C. A. Cardona, and L. E. Rincón, "Design and analysis of a second and third generation biorefinery: The case of castorbean and microalgae," *Bioresour. Technol.*, vol. 198, pp. 836–843, Dec. 2015.

23. J. M. Naranjo, C. A. Cardona, and J. C. Higuita, "Use of residual banana for polyhydroxybutyrate (PHB) production: Case of study in an integrated biorefinery," *Waste Manag.*, vol. 34, pp. 2634–2640, 2014.

24. I. X. Cerón, J. C. Higuita, and C. A. Cardona, "Analysis of a biorefinery based on *Theobroma grandiflorum* (copoazu) fruit," *Biomass Convers. Biorefinery*, vol. 5, pp. 183–194, 2015.

25. J. A. Dávila, M. Rosenberg, and C. A. Cardona, "A biorefinery for efficient processing and utilization of spent pulp of Colombian Andes Berry (*Rubus glaucus* Benth.): Experimental, techno-economic and environmental assessment," *Bioresour. Technol.*, vol. 223, pp. 227–236, 2017.
26. D. Lovarelli, J. Bacenetti, and M. Fiala, "Water footprint of crop productions: A review," *Science of the Total Environment*, vol. 548–549, pp. 236–251, 2011.
27. A. Y. Hoekstra, A. K. Chapagain, M. M. Aldaya, and M. M. Mekonnen, *The Water Footprint Assessment Manual. Setting the Global Standard*. London: Earthscan, 2011, pp.1–228.
28. N. S. Mohammad Sabli, Z. Zainon Noor, K. A. Kanniah, S. N. Kamaruddin, and N. Mohamed Rusli, "Developing a methodology for water footprint of palm oil based on a methodological review," *J. Clean. Prod.*, vol. 146, pp. 173–180, 2017.
29. D. Bocchiola, E. Nana, and A. Soncini, "Impact of climate change scenarios on crop yield and water footprint of maize in the Po valley of Italy," *Agric. Water Manag.*, vol. 116, pp. 50–61, 2013.
30. A. Y. Hoekstra, "The water footprint: Water in the supply chain," *Environmentalist*, vol. 93, pp. 12–13, 2010.
31. C. Liu, C. Kroeze, A. Y. Hoekstra, and W. Gerbens-Leenes, "Past and future trends in grey water footprints of anthropogenic nitrogen and phosphorus inputs to major world rivers," *Ecol. Indic.*, vol. 18, pp. 42–49, 2012.
32. M. M. Mekonnen and A. Y. Hoekstra, "Global gray water footprint and water pollution levels related to anthropogenic nitrogen loads to fresh water," *Environ. Sci. Technol.*, vol. 49, no. 21, pp. 12860–12868, 2015.
33. B. G. Subhadra and M. Edwards, "Coproduct market analysis and water footprint of simulated commercial algal biorefineries," *Appl. Energy*, vol. 88, pp. 3515–3523, 2011.
34. M. Wu and B. Sawyer, *Estimating Water Footprint and Managing Biorefinery Wastewater in the Production of Bio-based Renewable Diesel Blendstock*. Washington, DC: Energy Systems Division, Argonne National Laboratory, United States, 2016, pp. 1–26.
35. A. P. Mariano, M. O. S. Dias, T. L. Junqueira, M. P. Cunha, A. Bonomi, and R. M. Filho, "Butanol production in a first-generation Brazilian sugarcane biorefinery: Technical aspects and economics of greenfield projects," *Bioresour. Technol.*, vol. 135, pp. 316–323, 2013.
36. International Sustainability et Carbon Certification (ISCC 205), "GHG Emissions calculation methodology and GHG audit," 2011, V1.16, pp. 1–20.
37. US Environmental Protection Agency (EPA), "Greenhouse gas emissions," 2015. [Online]. Available: www.epa.gov/ghgemissions/overview-greenhouse-gases. Accessed October 2017.
38. G. J. Ruiz-Mercado, R. L. Smith, and M. A. Gonzalez, "Sustainability indicators for chemical processes: II. Data needs," *Ind. Eng. Chem. Res.*, vol. 51, no. 5, pp. 2329–2353, 2012.
39. J. Moncada, M. M. El-Halwagi, and C. A. Cardona, "Techno-economic analysis for a sugarcane biorefinery: Colombian case," *Bioresour. Technol.*, vol. 135, pp. 533–543, 2013.
40. S. Y. Pan, Y. J. Lin, S. W. Snyder, H. W. Ma, and P. C. Chiang, "Assessing the environmental impacts and water consumption of pretreatment and conditioning processes of corn stover hydrolysate liquor in biorefineries," *Energy*, vol. 116, pp. 436–444, 2016.

41. F. Cherubin and G. Jungmeier, "LCA of a biorefinery concept producing bio-ethanol, bioenergy, and chemicals from switchgrass," *Int. J. Life Cycle Assess.*, vol. 15, pp. 53–66, 2010.

42. R. Parajuli et al., "Environmental life cycle assessments of producing maize, grass-clover, ryegrass and winter wheat straw for biorefinery," *J. Clean. Prod.*, vol. 142, pp. 3859–3871, 2017.

43. H. L. Maranduba, S. Robra, I. A. Nascimento, R. S. da Cruz, L. B. Rodrigues, and J. A. de Almeida Neto, "Reducing the life cycle GHG emissions of micro-algal biodiesel through integration with ethanol production system," *Bioresour. Technol.*, vol. 194, pp. 21–27, 2015.

6

Social Analysis of Biorefineries

This chapter aims to present the first approaches to social assessment applied to biorefineries using mainly the Colombian case as an example, given the socioeconomic characteristics of this country. Throughout this chapter, the most relevant concepts to determine the social impact that generates this type of system are indicated: job generation at land and process levels, food supply, and health. However, it is not an easy task. Social analysis is mainly associated with the supply chain. People can expect that any efficient project related to a biorefinery can improve supply chains, making more profitable the use of biomass in an integrated way. Analyzing that separately can be a problem because impacts from stand-alone process are sometimes confused with those based on biorefineries. So, it is practically impossible to find specific data for the cultivation, harvesting, processing, and distribution when the evolution from one to many product systems is implemented because the rate of biomass use can be the same.

In Poland, the Institute for Fuels and Renewable Energy—IPIEO conducted a project to analyze demo and commercial-scale biorefineries [1] and summarized the social impact of the biorefineries in Europe as follows:

- Stabilization and provision of new employment opportunities.
- Revitalize existing industries.
- Promote regional development, especially in the R&D area.

These results can also be extended to other biorefineries in America, Africa, and Asia, where additionally to that, it is observed an increase of incomes in some regions mainly close to sugarcane-processing biorefineries. However, it is still in the crops production where the highest impacts are observed.

6.1 Job Generation

For the social assessment of biorefineries, there are different impacts that can be measured. The first one will be the impact associated with the number of generated employments and the second one, the general impact over the population linked to the process (local communities). Regarding the number of generated employments, there are associated indirect and induced jobs related to the supply chains. This job generation can be determined at two levels, at land and process levels.

6.1.1 Land Level

According to the Pilot Testing of GBEP Sustainability Indicators for Bioenergy in Colombia of the FAO [2], the total number of jobs in the sugarcane sector for the year 2010 was around 32,000, of which 25,500 were associated with feedstock production. It was estimated that 28.4 jobs were created (jobs in the agricultural, manufacturing and transport, communal, social, and personal services) for each direct job created in the processing stage. Another representative sector can be the oil palm for the production of biodiesel. For this case, it is reported an average of one worker every 3.2 hectares of crop [3]. Considering that the crop area of palm oil was around 960,000 hectares in 2015 [4], the generated field jobs were around 300,000. As discussed earlier, these crops are used today in a biorefinery way having best profits than before (when stand-alone processing was the main strategy). Finally, the quality of life is increased for all workers involved in the supply chain, not only for the field workers but also for the transporters and distributors for sugarcane and palm in countries like Colombia and Brazil [2–4].

In addition, the creation of jobs can benefit areas that have suffered the armed conflict in countries, for example, Colombia, boosting the development of rural regions that were previously unproductive due to this situation. The government has searched for reparation and justice measures for the people that have been victims of forced displacement, land loss, and massacres. Some examples of these measures are the Law of Justice and Peace, which stipulates that:

> "[...] the restoration programs directed to attending the human and social development of the victims, the communities and the offenders will be promoted, in order to re-establish the social links, which may comprehend, among others, actions directed to: f) promoting the entailment of the victims to productive processes and income generating programs, as well as vocational training programs, that may allow accessing to productive employments, stimulating the civil society and private sector support in order to facilitate the social reinsertion" [5].

It is really important to note that in countries where the illegal crops exist, the most effective strategy is to replace these crops with legal plantations contributing to farmers in the same way as illegal crops do. For example, in some places in Colombia where Marijuana was historically cultivated in 2015, avocado began to be an alternative contributing with more incomes, making today these regions free of illegal crops [6]. Additionally, the avocado not exported is used for other products including mashed avocado for food "guacamole". A proposal for a biorefinery for this surplus of avocado has been developed last year [7].

6.1.2 Process Level

According to the Pilot Testing of GBEP Sustainability Indicators for Bioenergy in Colombia of the FAO [2], the total number of jobs in the sugarcane sector was around 6,600 in the processing stage. These jobs were distributed between administrative and sales employees, professionals, technicians, and technologists. In addition, the inclusion of ethanol production to the sugarcane processing increases 22.3% of the jobs. According to Fedesarrollo [8], the total number of direct workers associated with ethanol production was between 6,500 and 7,500. Regarding oil palm crop, in Colombia, approximately 40.74% of the crude palm oil was used for biodiesel, implying that around 26,000 direct jobs could be attributed to biodiesel production [2].

6.2 Food Supply

Biorefineries, as facilities for the production of value-added products and energy carriers, can use first-, second-, or third-generation raw materials. It has generated different discussions related with their impact on environmental and social aspects. For this reason, different publications have been released to demonstrate the beneficial impact that the implementation of biorefineries could have from the environmental and social point of view through the applications of the so-called life cycle assessment (LCA) and social life cycle assessment (SLCA), respectively [9–12]. On the other hand, different discussions have aroused in terms of the use of water, land, and agricultural resources to produce biomass to supply the high demand of raw materials in the biorefineries. This fact also has generated a discussion between food production and the energy crops implementation. Therefore, an analysis of the impact of biorefineries in food supply (i.e., food security) should be performed.

The relation between biorefineries and food security has been a topic widely discussed in several reports and papers. In principle, this relation is based on the high agricultural products (e.g., oils, seeds) demand of the biorefineries, which affects directly the amount of production of food and non-food products. This can be reflected in the increment of the base product price that is common to the industrial and food sectors. For instance, the biodiesel production from edible oils such as palm oil reduces the amount of palm oil available to produce other food products such as olein, stearin, and margarines. This problem also has been analyzed from other perspective, the production of the raw materials that are necessary in a biorefinery competes directly with the resources used to produce food and non-food products in the same region (e.g., water, land, fertilizers, and so on) [13]. Therefore, an apparent reduction in the food and non-food products production from the agricultural sector as well as a high competence in the land use can be observed.

In this way, two main issues can be identified. The first one is the competition for agricultural products that can be used to produce bio-based and food products. The second one is the competition and stress generated in agricultural resources [14]. However, many approaches have been performed to demonstrate that the combination of biorefineries and food products from the same feedstock is possible without affecting the food security. [15–17]. As a result, the inclusion of biorefinery products in the productive chain of the agricultural feedstocks has been proposed. This inclusion involves the integration of technologies and conversion pathways as a linkage between the industrial and food sectors [18,19]. Thus, the use of agricultural feedstocks for the production of bio-based products, specially, energy carriers, could strengthen the links between the different industrial sectors and the agricultural market [20]. According to the abovementioned, the food security topic not only involves the use of first generation raw materials in a biorefinery system but also includes second- and third-generation feedstocks, attending to the second mentioned issue. Therefore, the idea of non-food competing feedstocks cannot be considered as the solution to improve or maintain the food security indicators calculated by the FAO (e.g., average value of food production, prevalence of severe food insecurity in the total population, per capita food production variability, access to improved water sources, and so on) [21,22]. Thus, the influence of the food security in the design of biorefineries should be evaluated using the three sustainability pillars (i.e., economy, social, and environmental impact).

Once the previous statements associated with food security and the use of agricultural feedstocks in a biorefinery system have been briefly discussed, the influence of the food security concept in the development of bio-based products can be performed. Maltsoglou et al. (2014) [20] evaluated the effect of the bioenergy production and its tightly relation with the food security concept. These authors studied the production of electricity in Malawi using the BEFS Rapid Appraisal methodology, which considers the following areas of analysis: (1) country context, (2) natural resources: biomass potential analysis, (3) techno-economic analysis, and (4) socio-economic analysis. This methodology was applied following the main items related with food security (i.e., availability, accessibility, stability, and use). Therefore, the analysis assessed the potential availability of biomass that is additional and/or complementary to food, feed, and other uses, and also the competition for resources and negative impacts on food production.

As a conclusion of the study developed by BEFS in FAO, the authors reported that the use of sunflower and soybean oils for electricity generation using the straight vegetable oil (SVO) as a fuel is the most cost-competitive option in comparison with other technologies such as biomass gasification using corncobs. Thus, the authors have found that the rural electricity production without affecting the food production and incomes from the food industry is possible in the Malawi context. Nevertheless, they suggest that there should be clear energy policies to ensure a good development of the non-interconnected electricity production with other well-established technologies such as diesel plants [20].

As can be seen in the above case of study, the analysis of the food security concept in the development or implementation of any bio-based product from a biotechnological or thermochemical pathway depends strongly on the social and political conditions of the country where the analysis is performed [23].

6.3 Health

In a biorefinery, health aspects are related with the end use of the obtained products [24,25]. These products can be focused on the food, pharmaceutical, and cosmetic industry. Compounds such as vitamins, carotenes, minerals, and polyphenolic compounds have a special effect on human health due to the protective effects against diabetes [25], neurodegenerative diseases [26], inflammatory processes [27], and cancer [28–30], among others. Phenolic compounds are linked to antioxidant activity of plants, which inhibit oxidation reactions of free radicals [31].

The consumption of polyphenolic compounds has an inverse relation with the risk of some cancer types; this has been demonstrated by epidemiologic studies. Polyphenols can interrupt some stages of cancer process since they possess anticarcinogenic properties and antioxidant activity [32,33]. The consumption of some polyphenolic compounds in large amounts can be harmful, affecting the metabolism of lipids, proteins, and polysaccharides [34,35].

Plants and their fruits are the main containers of antioxidant and polyphenolic compounds, and these are present as complex mixtures [25]. These compounds have physiological and biochemical functions with benefits to human health. For example, in the past years the term "functional food" has caught the attention since they are foods that contain natural substances with biochemical functions [36]. The knowledge of the feed composition allows us to choose suitably the transformation processes in order to generate the most convenient products and the most favorable configuration of biorefinery. However, the nutritional content depends specifically on cultivation conditions and specific plant genotype [37].

The suitable examples of feedstocks with considerable bioactive content are tropical fruits. Pulp, peels, and seeds of fruits are potential sources of bioactive compounds. Mango is the main tropical fruit produced worldwide, followed by papaya, avocado, and pineapple and are known as "major tropical fruits" [38]. The remaining fruits are known as "minor tropical fruits". In 2009, the production of major tropical fruits was estimated in 82.2 million tons [39]. The high demand of natural products has been supported by the growing need of sustainable products with high content of natural and bioactive compounds. It is important to note that the abovementioned publications demonstrated that the economic as well as the environmental performance of the natural products increases with the use of biorefineries, making this case a real example of sustainability when the advantages of using these products for the society is considered.

References

1. "Institute for Fuels and Renewable Energy—IPIEO," 2018. [Online]. Available: www.ipieo.pl/ Accessed January 2018.
2. Food and Agriculture Organization of the United Nations (FAO), "Results of pilot-testing of GBEP sustainability indicators for bioenergy in Colombia -Indicator 12: Jobs in the bioenergy sector," in *Pilot Testing of GBEP Sustainability Indicators for Bioenergy in Colombia*, pp. 37–108. Rome: FAO Corporate Document Repository, 2014.
3. H. García Romero and L. Etter Calderón, Evaluacion de la Politica de Biocombustibles en Colombia. Bogotá, Colombia, 2012.
4. Ministerio de Agricultura y Desarrollo Rural, "Red de Información y Comunicación del Sector Agropecuario Colombiano," *Estadísticas - Agrícola*, 2018.
5. Congreso de la República de Colombia, *Diario Oficial 45.980- Ley de Justicia y Paz*. Bogotá, Colombia: Colombia.
6. M. Bristow, "With Marijuana Price Down 70%, Colombia Growers Are Bailing," 2015. [Online]. Available www.bloomberg.com/news/articles/2015-07-13/with-marijuana-prices-down-70-in-colombia-farmers-are-bailing Accessed March 2018.
7. J. A. Dávila, M. Rosenberg, E. Castro, and C. A. Cardona, "A model biorefinery for avocado (*Persea americana* mill.) processing," *Bioresour. Technol.*, vol. 243, pp. 17–29, 2017.
8. M. A. Arbeláez, A. Estacio, and M. Olivera, *Impacto socioeconómico del sector azucarero colombiano en la economía nacional y regional*. Bogotá, Colombia: Fundación para la Educación Superior y el Desarrollo (Fedesarrollo), 2010.
9. N. Hajjaji, S. Martinez, E. Trably, J. P. Steyer, and A. Helias, "Life cycle assessment of hydrogen production from biogas reforming," *Int. J. Hydrogen Energy*, vol. 41, no. 14, pp. 6064–6075, 2016.
10. F. Cherubini and S. Ulgiati, "Crop residues as raw materials for biorefinery systems—A LCA case study," *Appl. Energy*, vol. 87, pp. 47–57, 2010.
11. F. Cherubini and A. Strømman, "Life cycle assessment of bioenergy systems: State of the art and future challenges," *Bioresour. Technol.*, vol. 102, no. 2, pp. 437–451, 2011.
12. M. Finkbeiner, M. S. Jørgensen, and M. Z. Hauschild, "Defining the baseline in social life cycle assessment," *Int. J. Life Cycle Assess.*, vol. 15, no. 4, pp. 376–384, 2010.

13. J. C. Escobar, E. S. Lora, O. J. Venturini, E. E. Yáñez, E. F. Castillo, and O. Almazan, "Biofuels: Environment, technology and food security," *Renew. Sustain. Energy Rev.*, vol. 13, no. 6–7, pp. 1275–1287, 2009.

14. T. Koizumi, "Biofuels and food security," *Renew. Sustain. Energy Rev.*, vol. 52, pp. 829–841, 2015.

15. K. T. Lee and C. Ofori-Boateng, "Sustainability of biofuel production from oil palm biomass," *Green Energy Technol.*, vol. 138, 2013.

16. C. A. Cardona, C. E. Orrego, and L. F. Guitiérrez, *Biodiesel*, 1st ed. Manizales: Universidad Nacional de Colombia—Sede Manizales, Gobernación de Caldas, 2009.

17. C. A. Cardona and C. E. Orrego, *Catalytic Systems for Integral Transformations of Oil Plants Through Biorefinery Concept*. Manizales, Colombia: Universidad Nacional de Colombia, Manizales campus, 2013.

18. N. Qureshi, D. Hodge, and A. Vertes, *Biorefineries: Integrated Biochemical Processes for Liquid Biofuels*, 1st ed. Burlington, VT: Elsevier, 2014, pp. 1–296.

19. Ó. J. Sánchez and C. A. Cardona, "Conceptual design of cost-effective and environmentally-friendly configurations for fuel ethanol production from sugarcane by knowledge-based process synthesis," *Bioresour. Technol.*, vol. 104, pp. 305–314, 2012.

20. I. Maltsoglou et al., "Combining bioenergy and food security: An approach and rapid appraisal to guide bioenergy policy formulation," *Biomass and Bioenergy*, vol. 79, pp. 80–95, 2014.

21. I. Maltsoglou, T. Koizumi, and E. Felix, "The status of bioenergy development in developing countries," *Glob. Food Sec.*, vol. 2, no. 2, pp. 104–109, 2013.

22. Food and Agriculture Organization of the United Nations (FAO), "Biofuels: prospects, risks and opportunities," *FAO-Electronic Publ. Policy Support Branch*, 2008.

23. J. A. Quintero, C. A. Cardona, E. Felix, J. Moncada, Ó. J. Sánchez, and L. F. Gutiérrez, "Techno-economic analysis of bioethanol production in Africa: Tanzania case," *Energy*, vol. 48, no. 1, pp. 442–454, 2012.

24. A. A. González, "Design and assessment of high technology processes for enhancing the viability of agribusiness based on the sustainable use of biomass in Amazonas," Universidad Nacional de Colombia. Departamento de Ingeniería Eléctrica, Electrónica y Computación., 2015.

25. I. X. Cerón Salazar, "Design and evaluation of processes to obtain antioxidant-rich extracts from tropical fruits cultivated in Amazon, Caldas and Northern Tolima regions," Ph Thesis. Universidad Nacional de Colombia. Departamento de Ingeniería Eléctrica, Electrónica y Computación, 2013.

26. R. ArunaDevi et al., "Neuroprotective effect of 5,7,3′,4′,5′-pentahydroxy dihdroflavanol-3-O-(2″-O-galloyl)-β-d-glucopyranoside, a polyphenolic compound in focal cerebral ischemia in rat," *Eur. J. Pharmacol.*, vol. 626, no. 2–3, pp. 205–212, 2010.

27. S. J. Wu et al., "Supercritical carbon dioxide extract exhibits enhanced antioxidant and anti-inflammatory activities of *Physalis peruviana*," *J. Ethnopharmacol.*, vol. 108, no. 3, pp. 407–413, 2006.

28. P. Kris-Etherton et al., "Bioactive compounds in foods: their role in the prevention of cardiovascular disease and cancer," *Am. J. Med.*, vol. 113 Suppl, no. 1, pp. 71S–88S, 2002.

29. A. R. Collins, "Antioxidant intervention as a route to cancer prevention," *Eur. J. Cancer*, vol. 41, no. 13, pp. 1923–1930, 2005.

30. G. M. Forster et al., "Rice varietal differences in bioactive bran components for inhibition of colorectal cancer cell growth," *Food Chem.*, vol. 141, no. 2, pp. 1545–1552, 2013.

31. M. C. Kou, J. H. Yen, J. T. Hong, C. L. Wang, C. W. Lin, and M. J. Wu, "Cyphomandra betacea Sendt. phenolics protect LDL from oxidation and PC12 cells from oxidative stress," *LWT—Food Sci. Technol.*, vol. 42, no. 2, pp. 458–463, 2009.

32. S. J. Wu, L. T. Ng, D. L. Lin, S. N. Huang, S. S. Wang, and C. C. Lin, "Physalis peruviana extract induces apoptosis in human Hep G2 cells through CD95/CD95L system and the mitochondrial signaling transduction pathway," *Cancer Lett.*, vol. 215, no. 2, pp. 199–208, 2004.

33. E. G. de Mejía, S. Chandra, M. V. Ramírez-Mares, and W. Wang, "Catalytic inhibition of human DNA topoisomerase by phenolic compounds in *Ardisia compressa* extracts and their effect on human colon cancer cells," *Food Chem. Toxicol.*, vol. 44, no. 8, pp. 1191–1203, 2006.

34. C. Vasco, J. Ruales, and A. Kamal-Eldin, "Total phenolic compounds and antioxidant capacities of major fruits from Ecuador," *Food Chem.*, vol. 111, no. 4, pp. 816–823, 2008.

35. C. Vasco, J. Avila, J. Ruales, U. Svanberg, and A. Kamal-Eldin, "Physical and chemical characteristics of golden-yellow and purple-red varieties of tamarillo fruit (*Solanum betaceum* Cav.)," *Int. J. Food Sci. Nutr.*, vol. 60, no. SUPPL. 7, pp. 278–288, 2009.

36. American Dietetic Association, "Position of the American dietetic association: Functional foods," *J. Am. Diet. Assoc.*, vol. 99, no. 10. pp. 1278–1285, 1999.

37. J. Scalzo, A. Politi, N. Pellegrini, B. Mezzetti, and M. Battino, "Plant genotype affects total antioxidant capacity and phenolic contents in fruit," *Nutrition*, vol. 21, pp. 207–213, 2005.

38. Food and Agriculture Organization of the United Nations (FAO), "Tropical fruits. Medium-term prospects for agricultural commodities. Projections to the year 2010," 2003.

39. Food and Agriculture Organization of the United Nations (FAO), "Intergovernmental group on bananas and tropical fruits. Current situation and short-term outlook." pp. 1–6, 2011.

7

Development of Biorefinery Systems: From Biofuel Upgrading to Multiproduct Portfolios

This chapter aims to indicate some specific examples of biorefinery systems, considering a logic result of evolution from main biofuel factories to multiproduct complex biorefineries. Some of them really exist or just are planned or proposed. The analyzed cases are sugarcane, corn, oil palm, lignocellulosic biomass, tropical fruits, and microalgae biorefineries. A detailed explanation of each system is presented, describing the transformation routes, platforms, and products. Each case study highlights a particular valorization and integration strategy according to the feedstock and target products. Finally, this chapter presents a general perspective of logistic concepts applied to biomass supply systems.

7.1 Sugarcane

Sugarcane is typically the feedstock used in the classic sugar-refining case, which has been used in countries like Brazil, Colombia, and India for a number of decades to produce sugar, bioethanol, heat, and power [1]. Historically, this raw material was used for food and beverage purposes, but in the past years the upgraded fuel ethanol was involved as the solution to sugars uncertainties in the market. The configuration of this biorefinery comprises four main processes: (i) sugar production, (ii) bioethanol production, (iii) cogeneration system, and (iv) wastewater treatment, and this biorefinery corresponds to Scenario 1 [1]. Figure 7.1 indicates the generalized diagram of a sugarcane biorefinery. The process begins with the reception of sugarcane for sugar production. The production process of sugar involves the following four main steps: cane milling, juice clarification, evaporation, and crystallization [2]. Molasses (a byproduct of sugar production) and in some cases the juice are used as the raw material in the production of fuel ethanol. This process includes four main steps: fermentation, distillation, dehydration, and stillage concentration [3]. The fermentation step is carried out mainly using *Saccharomyces cerevisiae* as the fermenting microorganism [4].

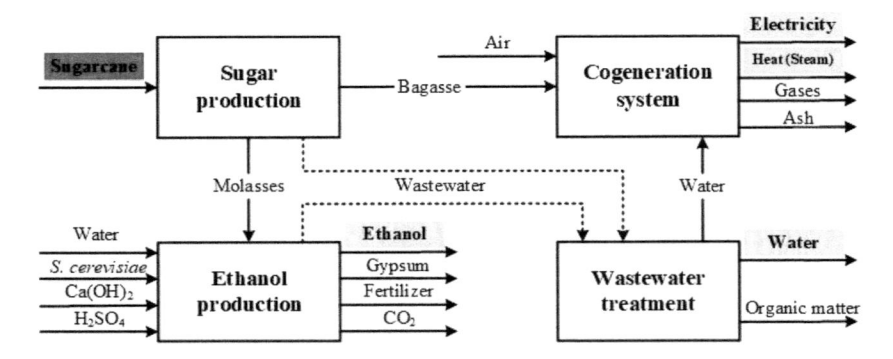

FIGURE 7.1
Diagram of classic sugarcane biorefinery.

The milling process also produces cane bagasse as a byproduct, which feeds the cogeneration system to produce heat and power. The most developed technology for the cogeneration system is the biomass integrated gasification combined cycle (BIGCC) [5], [6]. The cogeneration system comprises different stages such as biomass drying, biomass gasification, gas turbine, heat recovery steam generator (HRSG), and steam turbine [6]. It is very important to consider that mid-pressure and low-pressure steams are used to meet the heating requirements of the entire biorefinery system. Depending on the scale of production, the electricity needs of the biorefinery are fulfilled by the power produced from bagasse. High-pressure steam is used to generate power in the steam turbine. The used water is also treated with caustic soda and recycled to generate steam.

Finally, the wastewater treatment section comprises cooling and sedimentation tanks, and reverse osmosis modules to purify and separate the organic and inorganic fractions present in water [7]. It is very important to note that the purified water is recycled to meet the water demands of both the sugar and ethanol processes. The additional water is discharged arbitrarily after treatment. The organic/inorganic matter remaining after these processes can be collected and used as fertilizer and thus is a secondary product of the biorefinery [7].

Another use of this feedstock in biorefinery is for the additional obtaining of products anthocyanins and polyhydroxybutyrate (PHB) in order to promote the maximum exploitation of the sugarcane value chain through a very high value-added products [1]. The inclusion of PHB as a new product is promising because the demand for this biopolymer in the past years is increasing dramatically. In this biorefinery, sugar, bioethanol, heat, and power are obtained in the same way that the base case. This biorefinery corresponds to Scenario 3 [1]. Figure 7.2 shows the generalized diagram of alternative sugarcane biorefinery. PHB production is carried out using part of the same detoxified hydrolysates rich in glucose, obtained in the hydrolysis stage

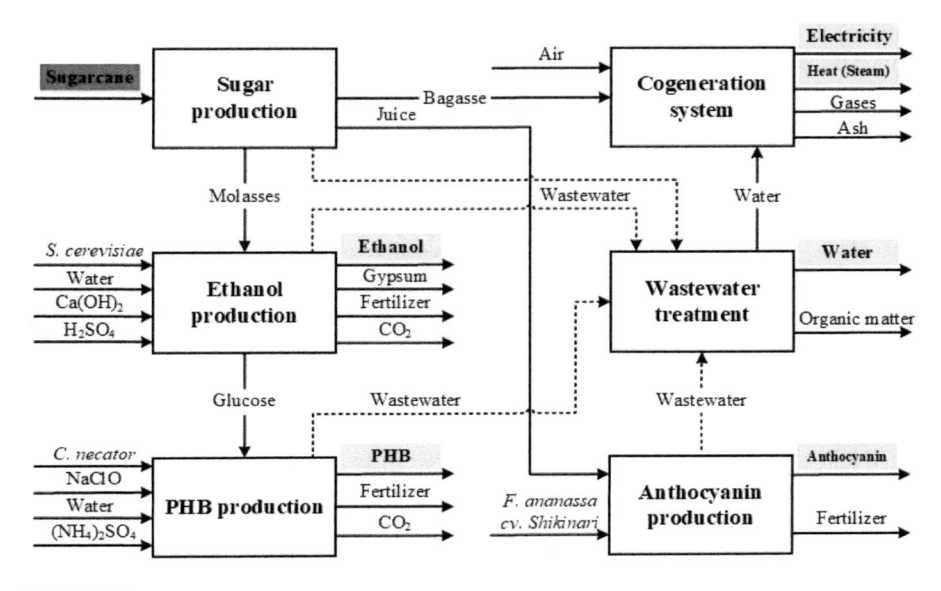

FIGURE 7.2
Diagram of alternative high value-added sugarcane biorefinery.

of the ethanol production process from molasses. The sugars are subjected to six steps: sterilization, fermentation using *Cupriavidus necator* (*Ralstonia eutropha*) as a microorganism, digestion with sodium hypochlorite, centrifugation, washing, and evaporation [1,8].

The production of anthocyanins by plant cell cultures *Fragaria ananassa* cv. Shikinari uses both disaccharide and monosaccharide sugars present in cane juice [9]. The three steps for anthocyanin production are as follows: (1) cell growth and cell pigmentation, (2) cell disruption and recovery of the metabolite, and (3) concentration and purification of the metabolite. The information about the technical, energy, and economic aspects of sugarcane biorefineries is presented in detail in Tables 7.1–7.3.

Steam generation is an important aspect to supply own biorefinery energy needs, considering that most of the water fed into the cogeneration system is

TABLE 7.1

Steam Characteristics Generated in Cogeneration Systems

Steam characteristics	Pressure (bar)	Temperature (°C)	Production (ton/day)	
			Scenario 1	Scenario 3
Low pressure	3	153.60	909.84	702.74
Medium pressure	30	253.94	1822.41	1407.59
High pressure	105	329.40	1370.23	1058.33

Source: Taken from Ref. [1].

TABLE 7.2

Production Capacities and Yields for Each Process in the Biorefinery

Process	Scenario	Production		Yield		Product Composition	
		Unit	Value	Unit	Value	Unit	Value
Sugar	1	Ton/day	538.39	Ton sugar/ton	0.11	% (wt)	98.5
	3		469.41	cane	0.10	sucrose	98.7
Ethanol	1	Liter/day	125734	L ethanol/ton	26.19	% (wt)	99.7
	Sc. 3		316654	cane	65.97	ethanol	99.8
Anthocyanin	Sc. 1	Ton/day	0.00	Kg	0.00	% (wt)	99.9
	Sc. 3		3.93	anthocyanin/	0.82	anthocyanin	99.9
				ton cane		mixture	
PHB	Sc. 1	Ton/day	0.00	Kg PHB/ton	0.00	% (wt) PHB	99.5
	Sc. 3		44.35	cane	9.24		99.6
Electricity	Sc. 1	MW	30.60	MJ/ton cane	550.76	–	–
	Sc. 3		21.75		391.43	–	–

Source: Taken from Ref. [1].

TABLE 7.3

Cost Estimation of Biorefinery Products When One Supervisor and Seven Operators Are Considered

Item	Cost and share (%)	Sugar		Ethanol		PHB (Scenario 3)	Anthocyanin (Scenario 3)
		Scenario 1	Scenario 3	Scenario 1	Scenario 3		
Raw materials	Cost	0.23	0.53	0.91	0.20	0.58	2.60
	Share	73.86	78.31	72.06	48.03	27.35	15.46
Operating labor	Cost	0.0004	0.0003	0.0018	0.0005	0.0035	0.0393
	Share	0.14	0.05	0.14	0.12	0.16	0.23
Utilities	Cost	0.00	0.08	0.02	0.12	0.88	6.61
	Share	1.28	12.34	1.37	29.55	41.32	39.31
Operating charges, plant overhead, maintenance	Cost	0.01	0.00	0.02	0.01	0.05	0.55
	Share	1.80	0.74	1.93	1.76	2.46	3.24
General and administrative cost	Cost	0.06	0.04	0.26	0.07	0.47	5.39
	Share	18.94	6.63	20.25	15.98	22.21	32.03
Capital depreciation[a]	Cost	0.01	0.01	0.05	0.02	0.14	1.64
	Share	3.98	1.94	4.25	4.64	6.49	9.73
Total	Cost	0.32	0.67	1.27	0.42	2.12	16.82
	Share	100.00	100.00	100.00	100.00	100.00	100.00
Sale price/total production cost	–	1.33	0.63	0.98	2.98	1.47	4.16

Source: Taken from Ref. [1].
[a] Calculated using the straight line method.

recovered from other processes. Table 7.1 shows steam-generated capacities for each scenario. Table 7.2 shows production capacities and yields, taking into account the distribution for raw materials mentioned in the scenarios description. Another significant aspect to include in the biorefinery analysis is the equivalent carbon dioxide as a measure of greenhouse gas (GHG) emissions. The GHG emissions represented as equivalent kilograms of carbon dioxide per kilogram of processed cane were 0.52 and 0.78 for the first and the third scenario, respectively [1]. Distribution and technologies included in the chosen products directly affect yields and the production costs. The operating cost includes various aspects inherent to the production process, such as raw materials, utilities, labor and maintenance, general plant costs, and general administrative costs. Annualized capital costs are also included [1]. Table 7.3 shows the cost estimation for each product considered in scenarios and the production cost per product for a determined social setting.

7.2 Corn

The cornstarch is a common feedstock used for ethanol production. Remaining components from this process, such as protein, fat, and fibers present in the corn, when mixed, are known as distillers dried grains with solubles (DDGS). This case study proposes the exploitation of DDGS as a feedstock for the production hydrocarbons due to its high-energy content. Previous studies have shown the potential of DDGS for the production of syngas by gasification, bio-oil by pyrolysis, and hydrocarbons by catalytic pyrolysis [10].

Biorefinery corn considers ethanol, BTX, and steam production using corn and DDGS as feedstocks [10]. Figure 7.3 indicates the flowsheet of corn biorefinery. Ethanol production is a classic "dry mill corn ethanol production", which includes pretreatment, enzymatic hydrolysis, fermentation, ethanol purification (distillation and molecular sieves), and DDGS separation. DDGS are pretreated using drying (moisture 2% wt) and milling. These DDGS can be used for feed or for thermochemical transformations. Pretreated feedstock is subjected to a catalytic pyrolysis. The obtained product is composed of a mixture of aromatics, gases, water, and nitrogen that is conducted to a collection system (two condensers and an electrostatic precipitator) in order to collect the bio-oil (in aqueous and oil phases). Aqueous phase is treated as wastewater and the oil phase is separated in light and heavy fractions. The light fraction contains benzene, toluene, and *p*-xylene that are recovered as individual compounds. The heavy fraction includes aromatic compounds that are subjected to hydroprocessing to obtain hydrocarbons. The gases obtained in catalytic pyrolysis are olefins, CO, and CO_2 that are separated by cryogenic separation. The stream of gases is sent to a boiler for combustion and steam is generated to supply the energy requirements of the plant [10]. Table 7.4 shows the production capacity and energy consumption of the corn biorefinery.

As previously noted, ethanol is the largest product in both mass and energy. However, BTX is the main product, which represents 40% of the total energy of the coproduct flow. On the other hand, the main olefinic products are ethylene and propylene, which represent the third main energetic vector of biorefinery. Figures 7.4 and 7.5 detail the best economic analysis result for the biorefinery.

The investment capital corresponds to the sum of installed capital cost, cost of working capital, total indirect cost, contingency of the project, and land use. The equipment cost corresponds to the largest investment that must be made. In total, the investment capital is 343.2 million USD. In the annual operation cost, the feedstock represents the largest contributor followed by catalysts and chemicals. The annual operation cost is $ 2.27 million USD.

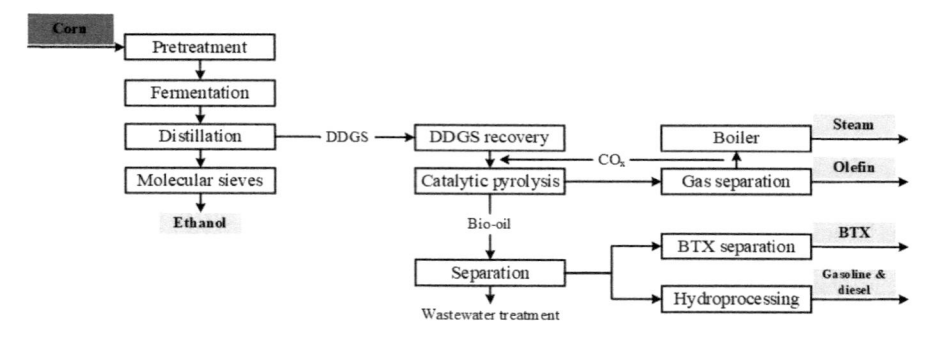

FIGURE 7.3
Scheme of corn biorefinery. (Adapted from [10].)

TABLE 7.4

Production Capacities for Each Process that Conforms the Biorefinery (Case Integrated)

Materials	Mass (Ton/day)	HHV (MJ/kg)	Energy (GJ/h)	Values ($MM/year)
Corn	2000	19.1	1588	208
Ethanol	–	29.8	1050	N.A.
BTX	131	41.8	228	55.5
DDGS	N.A	N.A	N.A	N.A.
Olefins	65	47.2	128	27.4
Gasoline	43.1	47.3	84.9	14.2
Diesel	37.8	44.8	70.5	12.5

Source: Adapted from Ref. [10].
N.A. No Apply.

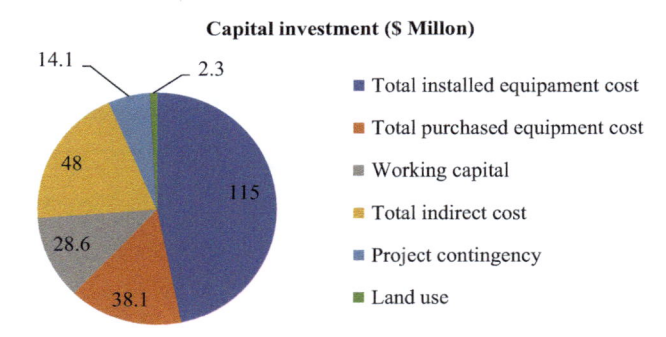

FIGURE 7.4
Distribution of capital investment. (Adapted from [10].)

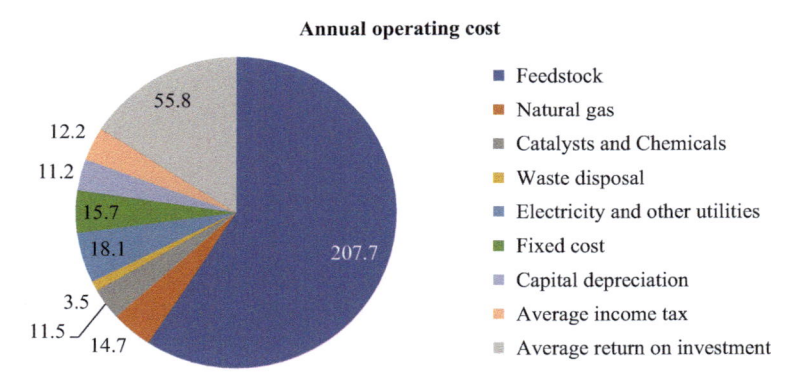

FIGURE 7.5
Distribution of annual operating cost. (Adapted from [10].)

7.3 Oil Palm

Palm is one of the most important oleo-chemical feedstocks in Indonesia, Malaysia, and Colombia. Currently, palm is used for biodiesel production, and glycerol is the main by-product obtained from this process. Additionally, the processing of palm for oil extraction leads to the formation of several residues such as empty fruit bunches (EFB). In this sense, the configuration of this biorefinery comprises four main processes: (i) biodiesel production, (ii) glycerol purification, (iii) ethanol production, and (iv) PHB production corresponding to Scenario 3. Scenario 4 considers the production of biodiesel, fuel ethanol, and PHB with mass integration of materials and recovery of waste streams [11]. Figure 7.6 indicates the generalized diagram of oil palm biorefinery. The biodiesel production can be described by three main sequential stages: esterification reaction, trans-esterification reaction, and distillation or

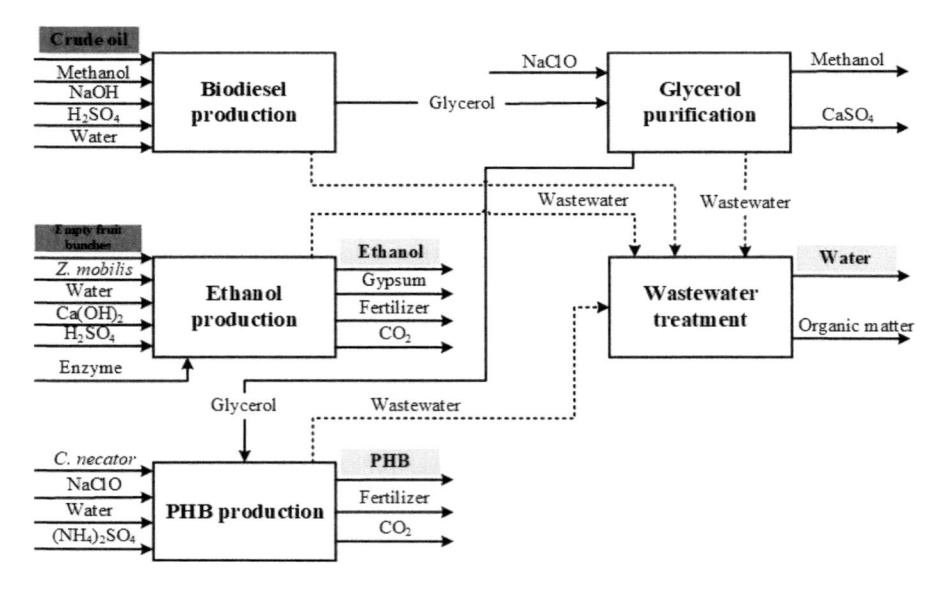

FIGURE 7.6
Diagram of oil palm biorefinery to produce biodiesel, ethanol, and PHB.

vacuum flash. Raw glycerol, coming from the biodiesel process, can be used as a carbon source to produce PHB. The glycerol is purified by removing and neutralizing the remaining methanol and the catalyst for obtaining a glycerol stream at high purity. The glycerol obtained in this process is used as a substrate for the PHB production by fermentation using *C. necator (R. eutropha)* as a microorganism [8]. Another product of this biorefinery is lignocellulosic ethanol, which uses EFB as a feedstock. When this type of raw material is used, the process to produce fuel ethanol can be described in six stages: dilute-acid hydrolysis, detoxification, enzymatic hydrolysis, fermentation, distillation, and dehydration [11], [12]. The fermentation step is carried out using *Zymomonas mobilis* as the fermenting microorganism [13].

Technical, economic, and environmental results obtained in this case study are presented later. Table 7.5 shows the material balance and energy requirements for Scenarios 3 and 4. The economic analysis of each scenario is presented in Table 7.6. The GHG emissions represented as equivalent carbon dioxide in ton per cubic meter of biodiesel are 0.54 and 0.51 for Scenarios 3 and 4, respectively [11].

Figure 7.7 shows a biorefinery scheme, where palm olein, biodiesel, methanol, heat, and power are generated using palm oil as a feedstock [14]. Crude palm oil (CPO) is obtained by extraction from oil palm. The fractionation of CPO is carried out by crystallization at low temperature. The olein fraction can be used in the food industry because it conserves 80–90% of carotenoids present in CPO and a minimum percentage of free fatty acids (FFAs) (0.1%).

TABLE 7.5

Main Process Streams and Energy Requirements for Scenarios 3 and 4 Studied for an Oil Palm-Based Biorefinery

Reagents/products	Scenario 3	Scenario 4
Reagents (kg/h)		
Crude oil	1000.0	1000.0
FFA content (%wt)	6.0	6.0
Methanol	160.0	111.5
H_2O	1877.0	841.7
H_2SO_4	65.0	15.0
EFB (moisture content 50%wt)	900.0	900.0
Products (kg/h)		
Biodiesel @ >99 wt%	1007.0	1007.0
Glycerol (intermediate) @ >99 wt%	85.8	85.8
Waste water	1314.5	9.5
Na_2SO_4	10.5	10.5
Ethanol @ >99.8 wt%	150.2	150.2
PHB @ >98 wt%	26.2	26.2
Energy requirements (GJ/ton)[a]		
Heating[b]	5.67	5.35
Cooling[c]	4.87	4.83
Electricity	2.16	2.19

Source: Taken from Ref. [11].
[a] Expressed in GJ per ton of processed feedstock.
[b] Heating utilities covered by low-pressure steam.
[c] Cooling utilities covered by cooling water.

The stearin fraction obtained in this stage is sent to biodiesel production, where sodium hydroxide is used as a catalyst. In this process, glycerol is obtained as a byproduct. The solid residues generated in CPO extraction are known as EFB and palm press fiber (PPF), which are used as feedstock to produce heat, power, and syngas through a BIGCC system. Furthermore, syngas is used as a platform to produce methanol using an isothermal Lurgi reactor. A percentage of methanol is sent to biodiesel production, and the remaining is sold. The gases go to a gas turbine to be burned with compressed air in order to generate electricity. The resulting stream in this process has low pressure and temperature and is used for steam generation through a HRSG system [14]. Technical, economic, and environmental results related to the production rates, yields, distribution costs, and potential environmental impact (PEI) are summarized in Table 7.7.

Another example of a biorefinery using oil palm as a feedstock is that described by Botero et al. (2017) to produce ethanol, biodiesel, xylitol, syngas, and electricity [15]. The environmental analysis was carried out using the Waste Reduction (WAR) Algorithm, developed by the Environmental

TABLE 7.6

Production Costs of Biorefinery Based on Oil Palm

	Scenario 3		Scenario 4	
Categories	Cost (USD/m³)	Share (%)	Cost (USD/m³)	Share (%)
Raw materials	478.97	61.12	421.50	57.79
Operating labor	20.97	2.68	20.97	2.87
Utilities	117.74	15.03	108.32	14.85
Operating charges, plant overhead, maintenance	38.09	4.86	38.09	5.22
General and administrative cost	63.64	8.12	63.64	8.73
Depreciation of capital	64.22	8.20	76.88	10.54
Subtotal	783.64	100.00	729.40	100.00
Credit by glycerol	0.00		0.00	–
Credit by ethanol	−223.21		−223.21	–
Credit by PHB	−93.22		−93.22	–
Total cost	467.21		412.97	–
Economic margin	58.65		63.45	–
Profitability index	1.36		1.38	–
Payout period (years)	4.65		3.57	–

Source: Taken from Ref. [11].

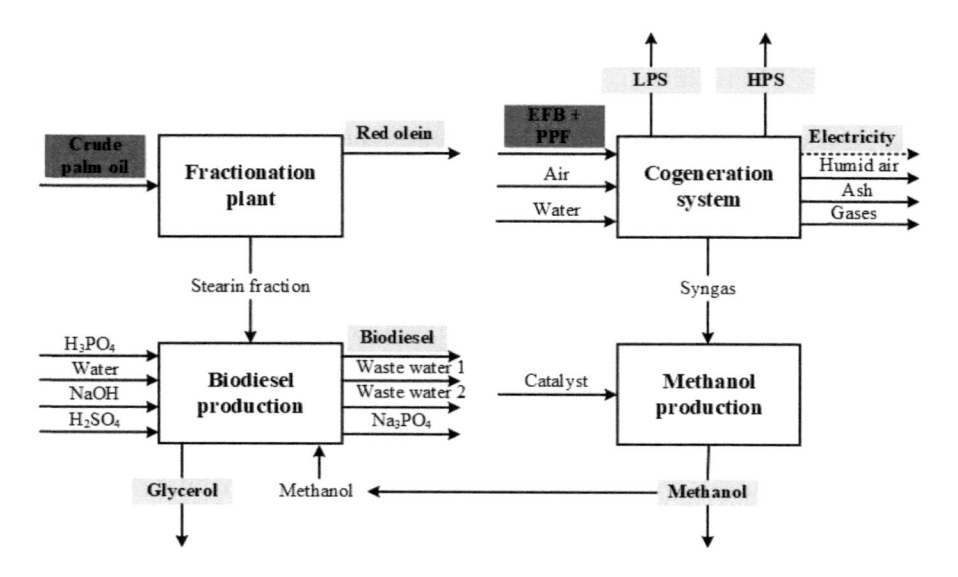

FIGURE 7.7
Diagram of oil palm biorefinery to produce biodiesel, palm olein, and methanol. (EPF: empty fruit bunches, PPF: palm press fiber, LPS: Low-pressure steam, HPS: high-pressure steam.)

TABLE 7.7

Main Simulation and Economic Results of Palm Oil Biorefinery

Reagents/Products	Kg/h
Reagents	
Crude palm oil	37,370.0
EFB+PPF	246,713.0
NaOH	35,638.0
Water	720,611.0
Air	2,000,000.0
H_2SO_4 @ >98 wt%	4,671.0
Energy	
Electricity generated	171.0
Heating	–
Cooling	215.0
Products	
Biodiesel @ >99 wt%	5,490.0
Glycerol UPS	616.0
Olein	31,973.0
Methanol	46,872.0
Residues	
Wastewater	5,509.0
NH_2SO_4	58.00
Flue gas	2,177,624.0
Ash, char, and tar	2,211,290.0
Share of production costs (%)	
Raw materials	53.84
Operation	18.91
Utilities	25.34
Capital depreciation	1.91
Total sales to total biorefinery costs ratio	1.16
Potential environmental impact per ton of products	240 using oil as fuel

Source: Taken from Ref. [14].

Protection Agency (EPA). The studied scenarios are as follows: Scenario 1, production of biodiesel, glycerol, xylitol, electricity, syngas, and ethanol; Scenario 2, production of biodiesel, xylitol, syngas, electricity, and ethanol; Scenario 3, production of biodiesel, syngas, electricity, and ethanol; and Scenario 4, biodiesel production. As can be seen in Figure 7.8, environmental impacts of the process diminished inversely proportional to the number of products.

On the other hand, Table 7.8 shows the price distribution for the biorefinery studied. When comparing the sale prices and the production costs,

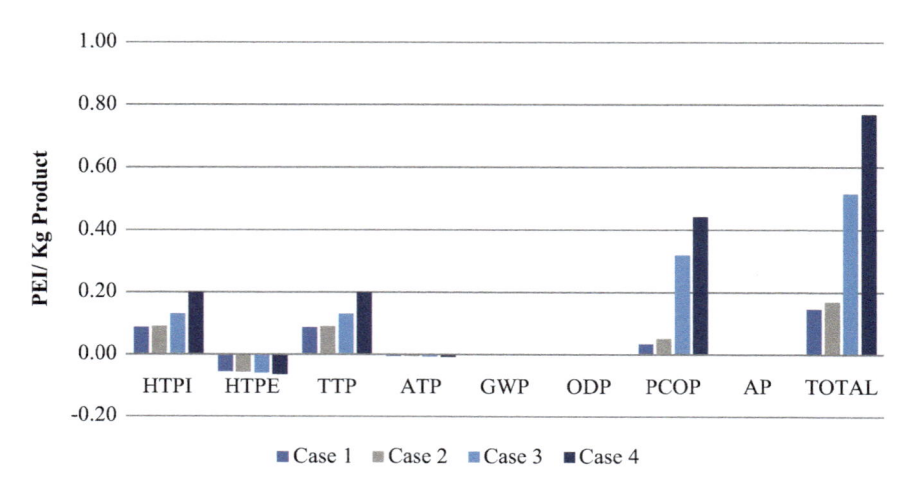

FIGURE 7.8
Results of environmental analysis. (Taken from Ref. [15].)

TABLE 7.8

Distribution of Production Costs

Product	Percent	Cost production		Sale Price		Reference
Biodiesel	17.97	0.03	USD/gal	3.64	USD/gal	[16]
Xylitol	29.44	15.66	USD/kg	164.00	USD/kg	[17]
Syngas	39.88	16.97	USD/m³	0.62	USD/m³	[18]
Glycerol	3.41	0.002	USD/L	53.24	USD/L	[19]
Electricity	9.30	4.95	USD/ kW	0.1	USD/ kW	[20]

it can be concluded that the biodiesel, xylitol, and glycerol had a profit margin higher than 99%, while the other products had production costs higher than sale prices.

Another biorefinery case using palm is the one performed by Aristizábal et al. (2016), where biodiesel, electricity, hydrogen, butanol, ethanol, and acetone are produced [12]. In this work, results of the potential environmental impact per kilogram of product are presented in Figure 7.9, considering each process stage and the global biorefinery.

Additionally, Figure 7.10 shows the percentage distribution of production costs, raw material costs, utility costs, and depreciation. As can be seen, operating costs contribute approximately 50% of the total production costs. Both the costs of raw materials and utilities contribute the least, this behavior is directly related with the integrated approach of the biorefinery.

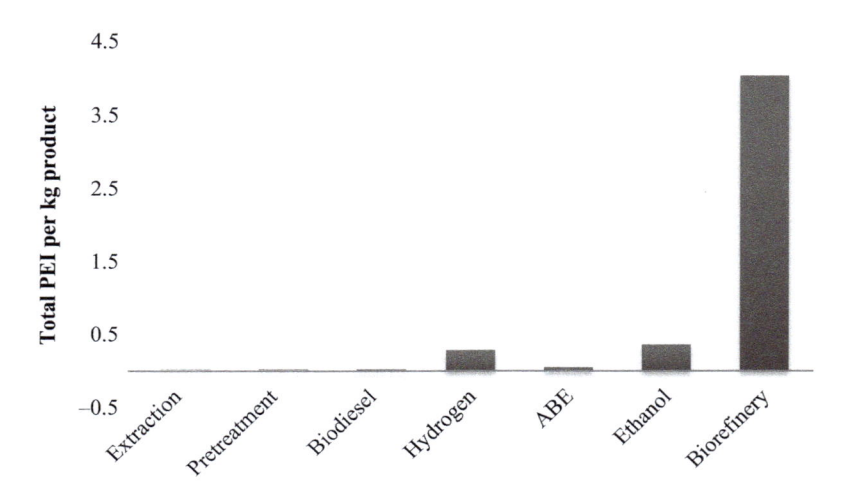

FIGURE 7.9
Potential environmental impacts for each biorefinery process. (Taken from Ref. [12].)

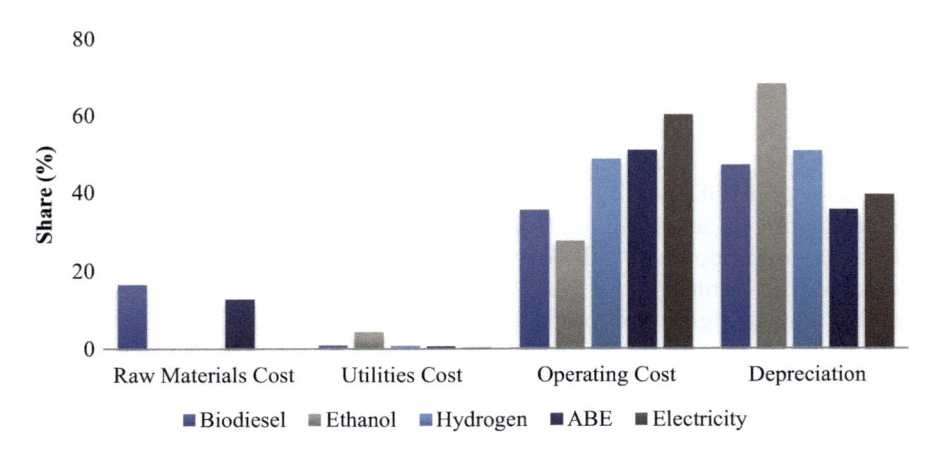

FIGURE 7.10
Share of distribution costs. (Taken from Ref. [12].)

Figure 7.11 shows the economic margins versus sale prices for products of the biorefinery. Due to a relatively high flow of biodiesel, its economic margin is positive for any sale price. For hydrogen, the flow is relatively low but it can also present a positive economic trend due to its high sale price. In conclusion, biorefinery presents a negative economic margin for minimum and maximum sale prices of the evaluated products due to the low productivity of the majority of the stages of the process.

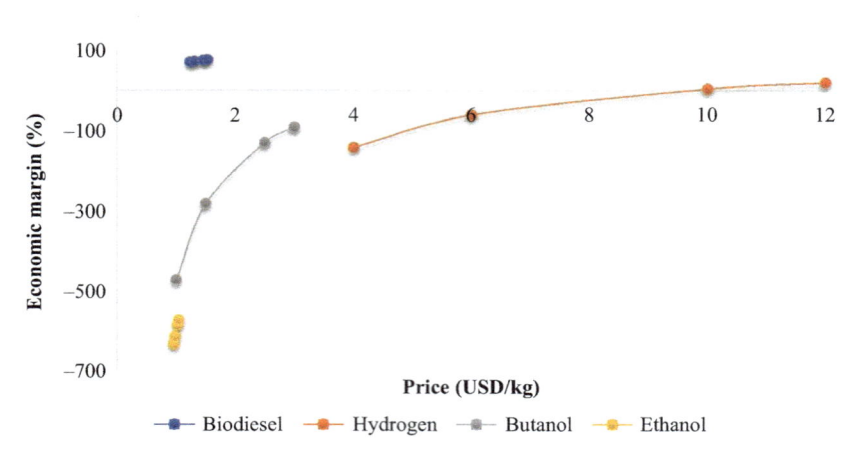

FIGURE 7.11
Economic margins for products of the biorefinery. (Taken from Ref. [12].)

7.4 Lignocellulosic Biomass

Lignocellulosic biomass is generally considered a promising feedstock due to its availability in large quantities, its relatively low cost, and its potential composition. Sugarcane bagasse (SCB) is a fibrous residue obtained after extracting the juice from sugarcane in the sugar production process. Coffee cut-stems (CCS) is a cut above the land, where the coffee plant is cultivated and obtained by crop renewal. Energy demand has generated a renewed interest in producing fuels from biomass. Biomass is made up of three major components: cellulose, hemicellulose, and lignin. Cellulose can be transformed into glucose, and then glucose is fermented to obtain ethanol. Ethanol is highly commercial and is used as a large-scale transportation fuel. Hemicellulose present in biomass can be broken into xylose, and then xylose is dehydrated to produce furfural. Furan-based compounds are highly versatile and key derivatives used in the manufacture of a wide range of important chemicals and can serve as precursors for the production of jet fuel substitutes or additives (alkanes C_7–C_{15}) [21].

According to the previous statements, the following biorefinery has a goal to transform lignocellulosic biomass in renewable fuels for the transport industry, both ground and air. The biorefinery considers the integrated production of ethanol, octane, and nonane as main products and building block products such as furfural and hydroxymethylfurfural (HMF; furan-based compounds) from SCB and CCS in Scenarios 3 and 4. (The same products are obtained in both scenarios but the distribution of sugars is different). This biorefinery considers six process stages: (i) sugars extraction, (ii) ethanol production, (iii) furfural production, (iv) octane production, (v) HMF production, and (vi) nonane production [22]. Figure 7.12 indicates the generalized

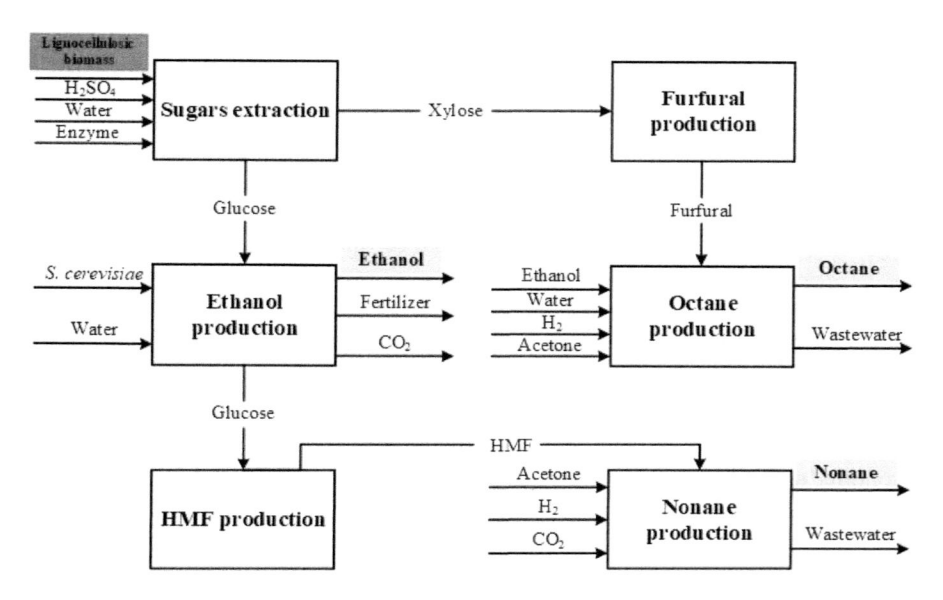

FIGURE 7.12
Diagram of sugarcane bagasse and coffee cut-stems biorefinery.

diagram of sugarcane and CCS biorefinery. In the extraction of sugars, ligno-cellulosic biomass is subjected to a process consisting of three stages: particle size reduction, dilute-acid pretreatment, and enzymatic hydrolysis in order to obtain xylose and glucose that are the platforms to generate furfural and HMF, respectively. Furan compounds are obtained from sugar-rich liquor by xylose or glucose dehydration. In the case of the glucose, this sugar is also used to produce ethanol by a fermentation process using *S. cerevisiae*. The purification step of ethanol is similar to the abovementioned cases.

Octane is obtained from furfural. The process to obtain octane is described in three steps: aldol-condensation reaction, hydrogenation reaction, and dehydration/hydrogenation reactions. Nonane is obtained from HMF, the process to obtain nonane is described in three steps: aldol-condensation reaction, hydrogenation reaction, and dehydrogenation/hydrogenation reactions. Both alkanes are purified using distillation process. The information about the technical, economic, and environmental aspects of sugarcane and CCS biorefineries is presented in detail in Tables 7.9 and 7.10.

Lignocellulosic biomass appears as an alternative to first-generation (1G) feedstocks (defined in Chapter 2) due to its availability in large quantities, relatively low cost, and a significant reduction of competition with food. Lignocellulosic biomass can be used for biofuels production such as bio-ethanol and biochemicals. However, the high degree of complexity is the main limiting factor because of its nature and composition (complex polymer composed of three carbohydrates: cellulose, hemicelluloses, and lignin). Cellulose and hemicellulose should be broken down into fermentable

TABLE 7.9

Production Capacities per Scenario and Production Cost of Biorefinery Products
for SCB and CCS

		Ethanol		Octane		Nonane		Furfural		HMF	
Scenario	Raw material	(kg/h)	(USD/ kg)	(kg/h)	(USD/ kg)	(kg/h)	(USD/ kg)	(kg/h)	(USD/ kg)	(kg/h)	(USD/ kg)
Sc. 3	SCB	310.24	0.84	644.18	2.63	398.85	10.27	681.52	0.47	597.27	0.35
	CCS	270.66	0.84	925.65	2.58	339.74	10.51	979.30	0.38	521.38	0.39
Sc. 4	SCB	366.01	1.83	–	–	664.66	10.22	–	–	995.45	0.32
	CCS	527.99	1.31	–	–	549.17	10.24	–	–	868.97	0.37

Source: Taken from Ref. [22].

TABLE 7.10

Economic and Environmental Results of Biorefineries from SCB and CCS

Scenarios	NPV (USD/period)	PEI/kg product	Economic margin (%)
SCB results			
Sc. 3	10,875,126.8	1.26	43.50
Sc. 4	–9,708,502.4	1.46	42.72
CCS results			
Sc. 3	31,805,819.8	2.10	30.43
Sc. 4	–7,536,606.5	2.20	33.31

Source: Taken from Ref. [22].

sugars in order to be converted into ethanol or value-added products. The
production of fermentable sugars is possible through additional process
stages called pretreatment and hydrolysis [23]. The inclusion of these stages
involves high operating and investment costs, which is the main reason why
ethanol from this feedstock has not made its breakthrough to industrial levels yet [24]. Figure 7.13 shows the ethanol production flowsheet when lignocellulosic biomass is considered as a feedstock.

The ethanolic fermentation process proceeds with *S. cerevisiae* or *Z. mobilis* when hexoses or hexoses plus pentoses are considered as a platform,
respectively. Initially, sugar-rich liquor is subjected to a sterilization process
at 121°C, in which the biological activity is neutralized. Later, the fermentation process is carried out using *S. cerevisiae* at 37°C or *Z. mobilis* at 30°C,
respectively. Afterward, cell biomass is separated from the culture broth by a
simple gravitational sedimentation technology. After the fermentation stage,
the culture broth containing approximately 5%–10% (wt) of ethanol is taken
to the separation step, which consists of two distillation columns. In the first
column, ethanol is concentrated nearly to 45%–50% by weight. In the second
column, the liquor is concentrated until the azeotropic point (96% wt) to be
led to the dehydration step with molecular sieves to obtain an ethanol concentration of 99.6% by weight.

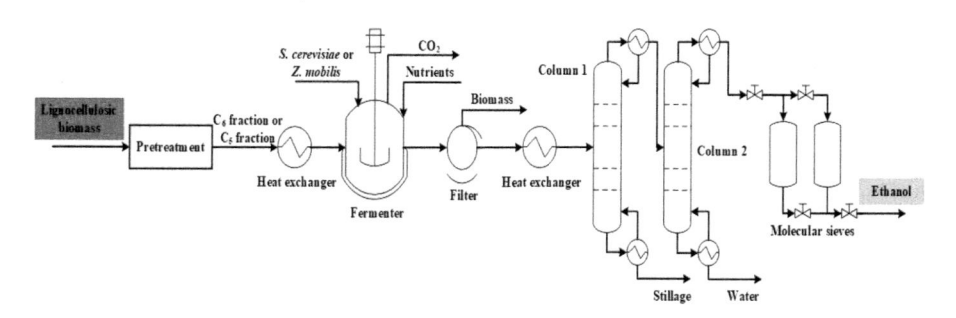

FIGURE 7.13
Scheme of ethanol production using lignocellulosic biomass.

In the pretreatment and enzymatic hydrolysis stages of lignocellulosic biomass, the lignin is obtained as a by-product as well as in the pulp industry as a coproduct. Its composition makes a potential source of products in food, textile, chemical, pharmaceutical, and cosmetic fields. The majority of lignin produced worldwide (95%) is used in the energy production through cogeneration systems, while the remaining 5% has been marketed for obtaining dispersants, surfactants, antioxidant, adhesives, and rubbers [25,26].

SCB, EFB, rice husk (RH), and CCS are examples of lignocellulosic feedstocks that can be used to produce ethanol [24]. Results of ethanol production costs for the evaluated crop residues are shown in Table 7.11. The highest production cost is obtained for SCB, while the lowest is obtained for EFB [24].

TABLE 7.11

Bioethanol Production Cost from SCB, EFB, RH, and CCS. Standalone Ethanol Plant

	SCB		EFB		RH		CCS	
Category	USD/L	Share (%)	USD/L	Share (%)	USD/L	Share (%)	USD/L	Share (%)
Raw materials [a]	0.3472	45.32	0.1948	33.71	0.1972	30.84	0.2387	35.06
Operating labor [b]	0.0037	0.48	0.0037	0.64	0.0037	0.58	0.0037	0.54
Utilities	0.2835	37.00	0.2639	45.67	0.3126	48.89	0.3098	45.52
Operating charges, plant overhead, maintenance	0.0126	1.65	0.0122	2.12	0.0130	2.04	0.0129	1.90
General and administrative cost	0.0518	6.76	0.0380	6.57	0.0421	6.59	0.0452	6.64
Depreciation of capital [c]	0.0674	8.80	0.0653	11.29	0.0708	11.07	0.0703	10.33
Total	0.7662	100.00	0.5779	100.00	0.6393	100.00	0.6807	100.00

[a] Raw material prices, SCB: US$15/ton, EFB: US$5/ton, RH: US$5/ton, CCS: US$18/ton.
[b] Used low pressure steam price was US$8.18/ton.
[c] Calculated using the straight line method.

7.5 Tropical Fruits

Citrus (orange and mandarin) is the most widely produced fruit in the world, and it is grown in more than 80 countries. Currently, citrus agroindustry is not a well-established chain, and it requires new opportunities to increase its valorization level. Citrus can be considered as an exceptional feedstock to the designing and assessing of biorefineries due to its versatile chemical composition. In this sense, this example proposes a citrus-based biorefinery for the integrated production of essential oil, concentrated juice, antioxidant, citrus seed oil, pectin, xylitol, PHB, ethanol, citric acid, lactic acid, and electricity [27].

Initially, the entire fruit is received, and a pulping process is carried out in order to separate the seeds, pulp, and peel. Once these three fractions are obtained, the pulp is used for the extraction of concentrated juice. The juice is a product, whereas the fiber stream is used as a raw material for pectin production.

On the other hand, the peel is sent to the production process of essential oil, where the volatile fraction present in the peel is extracted by supercritical fluid (CO_2). Until now, it is evident that the sequence showed preserves the characteristics and importance of these products, moreover the applications of these products in food and pharmaceutics industries require high purity and low degradability. On the other hand, the solid material remaining from the extraction of essential oil is rich in antioxidants.

Therefore, the remaining peel solids are mixed with the solid material from seeds to obtain antioxidants and oil from the seeds by supercritical fluid extraction (CO_2). The solid material resulting from last processes are still rich in polysaccharides such as pectin and lignocellulosic complex. Therefore, these characteristics are exploited in the extraction of pectin. Once the pectin is extracted, a solid material, liquor rich in polysaccharides, and soluble sugars are obtained. This stream is treated in the sugar extraction process using acid hydrolysis to produce xylose and glucose as platform products for obtaining xylitol, ethanol, PHB, lactic acid, and citric acid. The obtained xylose is sent to the xylitol production process and the liquor rich in glucose is divided as follows: 20% for ethanol, 20% for PHB, 30% for lactic acid, and 30% for citric acid. The production of xylitol is described in three steps: fermentation using *Candida guilliermondii* as a microorganism, crystallization, and centrifugation. In the case of lactic acid, the production is carried out through fermentation using *Lactobacillus delbrueckii*; and the citric acid is produced in three steps: fermentation using *Aspergillus niger*, precipitation, and crystallization. The production of ethanol, PHB, and electricity follows the same stages described in the abovementioned biorefineries. Figure 7.14 shows the simplified process block diagram for a citrus-based biorefinery.

Theobroma bicolor, known as Makambo, maraca, and cacao blanco is an Amazonian tree, which grows in different regions of Central and South

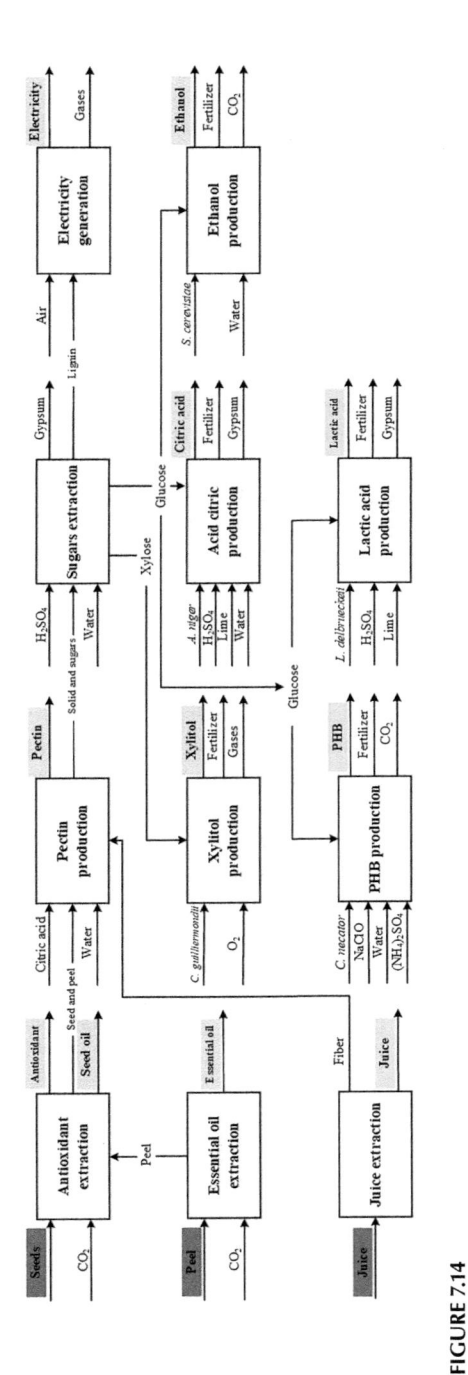

FIGURE 7.14
Diagram of orange and mandarin biorefinery.

America. Traditionally, its pulp has been used for direct consumption, whereas its seeds have been used as a kind of cocoa, and its peel has been generally disposed as waste. The biorefinery concept, for the joint production of bioenergy, biomaterials, biomolecules, and food products, has been adopted by different sectors in order to favor the integral exploitation of bio-based feedstocks and as a general rule for developing bio-based economy in small communities [28]. As an application to the general abovementioned remarks, the Makambo biorefinery is proposed in order to show the potential of tropical feedstocks as a source of value-added products.

Makambo biorefinery considers obtaining pasteurized pulp, butter, residual cake (paste used as a substitute for cacao), extract rich in phenolic compounds, biogas, and biofertilizer [29]. The first three products can be directly obtained in rural areas and do not require large investment and sophisticated technologies. However, this configuration generates a considerable amount of waste. For this reason, the biorefinery concept is implemented in order to increase the extent to which the entire fruit is used. The fruit is pulped and the seeds are mechanically extracted. Then, the pulp is homogenized, pasteurized, and frozen. After, the seeds are dried, and the butter and the residual cake are extracted by mechanical compression, the seeds are the base to obtain these products. The cake fraction obtained from the butter extraction and the peels are the platform to the antioxidants extraction. These components rich in phenolic compounds are extracted with supercritical carbon dioxide. Before the extraction, both, peels and cocoa are mixed, dried, and then milled. The carbon dioxide is recovered and recycled to the process after a purge. The extract is concentrated using the ultrafiltration technology. All solid residues that are obtained from the different processes are used to produce biogas and biofertilizer. These solids are mixed and subjected to anaerobic digestion, where biogas and slurry are the product streams. The slurry generated in the reactor is filtered, and the solid stream is the biofertilizer. The flowsheet for Makambo biorefinery is depicted in Figure 7.15.

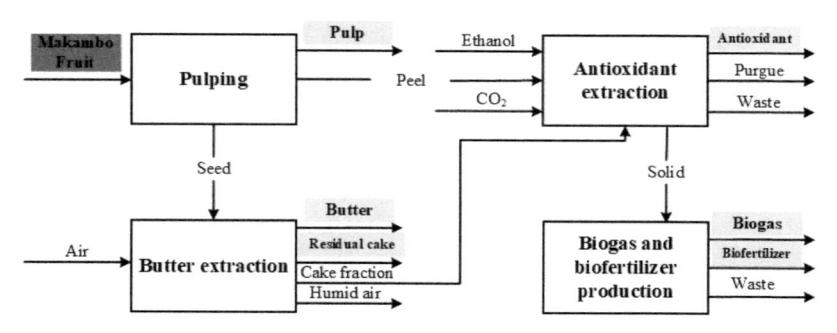

FIGURE 7.15
Diagram of Makambo biorefinery.

TABLE 7.12

Material Balances for Scenarios 1 and 2

Scenario	Material	Kg/day	Products	Kg/day	Residues	Kg/day
Sc. 1	Makambo fruit	1000.00	Pasteurized pulp	257.00	Wastewater	185.82
			Moisture (%)	87.90	Solid waste	382.99
			Residual cake	113.08		
			Moisture (%)	1.24		
			Lipid (%)	4.56		
			Butter	61.11		
			Moisture (%)	20.59		
			Lipid (%)	76.02		
Sc. 2	Makambo fruit	1000.00	Pasteurized pulp	257.00	Wastewater	619.44
	Ethanol (60%)	1.69	Moisture (%)	87.90	CO_2	39.87
	CO_2	40.20	Residual cake	56.54		
	Water	231.52	Moisture (%)	1.24		
			Lipid (%)	4.56		
			Butter	61.11		
			Moisture (%)	20.59		
			Lipid (%)	76.02		
			Phenolic comp. extract	5.36		
			Moisture (%); Phenolic compound (%)	45.83; 53.63		
			Biogas	142.35		
			CO_2 (%)	30		
			CH_4 (%)	70		
			Biofertilizer	91.35 •		

Source: Taken from Ref. [29].

Table 7.12 presents the mass balances for Scenarios 1 and 2 of Makambo biorefinery, respectively, and Table 7.13 shows the annualized costs including raw materials, operating labor, maintenance, utilities, operating charges, plant overhead, general and administrative and capital depreciation for each scenario. Additionally, revenues, profitability index, payout period, and net present value are also included.

TABLE 7.13

Annualized Costs and Economic Metrics for Each Scenario

Feature	Sc. 1		Sc. 2	
	USD/year	Share (%)	USD/year	Share (%)
Raw materials	102,270.00	16.49	118,048.00	12.47
Operating labor	87,660.00	14.13	93,883.90	9.92
Maintenance	7,516.85	1.21	18,518.20	1.96
Utilities	60,762.90	9.80	62,363.70	6.59
Operating charges	21,915.00	3.53	23,471.00	2.48
Plant overhead	47,588.40	7.67	56,201.00	5.94
General and administrative	26,217.10	4.23	29,798.90	3.15
Capital depreciation	266,339.00	42.94	544,000.00	57.49
Total cost	620,269.25	100.00	946,248.70	100.00
Revenues (USD/year)	1,141,960.00		2,701,880.00	
Profitability index	1.12		1.35	
Payout period (years)	7.33		5.11	
NPV (USD/year)	805,665.00		400,9910.00	

Source: Taken from Ref. [29].

7.6 Microalgae

Microalgae are very similar and the most simple form of plants. These can grow in different conditions and carry out the photosynthesis process as conventional plants in order to convert water, CO_2 (or other carbon sources), and sunlight into oxygen, lipids, and biomass [30]. Due to their simple cellular structure, it is possible to access easily to their nutrients, water, and CO_2 [31]. *Chlorella vulgaris, Spirulina maxima, Neochloris oleoabundans,* and *Dunaliella tertiolecta,* among others, are the most common species of microalgae and with higher oil content [32].

Biodiesel, glycerol, and energy (steam and electricity) are products obtained in microalgae biorefinery [32]. The flowsheet for microalgae biorefinery is shown in Figure 7.16. Biodiesel production contains three generalized sequential stages: pretreatment, reaction, and separation. In the pretreatment stage, particles and residues are removed from oil by filtration. Additionally, a pre-esterification reaction is carry out when the FFA content in the oil is >4%, in order to eliminate them. Then, the reaction stage involves a trans-esterification, where the triglycerides react with a short-chain alcohol to produce three molecules of alkyl esters and one of glycerol. The most common alcohol is methanol because of its properties and low cost. This reaction needs a catalyst that can be acid, basic, enzymatic, and heterogeneous. $NaOH$, KOH, H_2SO_4, HCl, zeolites, TiO_2, and lipases are examples of

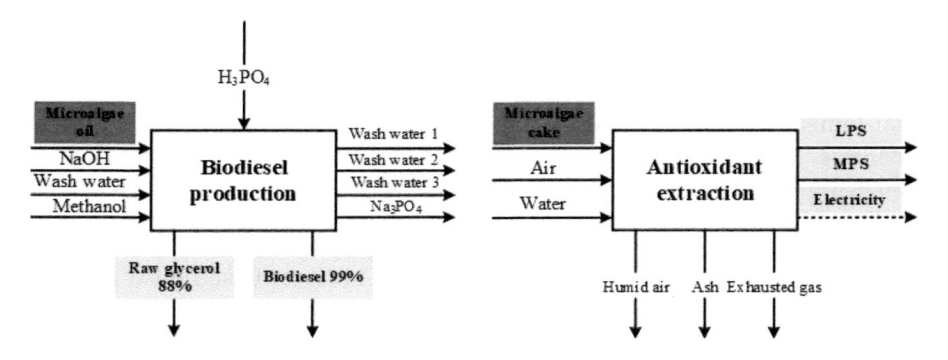

FIGURE 7.16
Diagram of microalgae biorefinery. (LPS: Low-pressure steam, MPS: Medium-pressure steam.)

catalysts. Finally, the separation stage is carried out following two options: (i) liquid–liquid extraction for recovering biodiesel in the light phase and glycerol in the heavy phase, and non-reacted alcohol is separated and (ii) vacuum distillation for separating non-reacted alcohol and liquid–liquid extraction to obtain glycerol and biodiesel. Then, it is water washed to achieve a high-purity biodiesel [32]. To the biodiesel process, a cogeneration system is added in order to generate electrical and thermal energy, improving process economy by reducing the need for external utilities and by selling generated electricity. The cogeneration system used in this biorefinery follows the same principle explained in sugarcane biorefinery (Section 7.1).

7.7 Influence of Logistics

The viability of biorefineries and bioprocesses are directly linked to supply logistics of biomass. The economic efficiency of supply logistics determines a promising biorefinery. The growing demand of biomass for the production of biofuels and biochemicals requires greater complexity in supply systems. These systems require exhaustive planning approaches, where the knowledge of production areas, consumers, biomass availability, and infrastructure are essential.

In terms of quality and quantity, biomass supply can vary significantly because of the factors such as production techniques, plant size, and facilities for conditioning, among others. In the case of high-scale biorefineries, the biomass supply should be continuous, in large quantities, and versatile using biomass from different nature. However, the versatility is a topic depending on constant biomass availability.

To establish supply systems according to production technology and regional conditions is the main challenge that faces the biomass logistics

[33]. Additionally, the biomass interchangeability, seasonal changes, market fluctuations, and existing structures for its generation and supply must also be considered [34]. For ensuring a suitable and feasible biomass supply, integral logistic systems are needed based on the requirements of all parties involved, producers, consumers, and society [33]. The biomass production and acquisition should be planned and organized, and the supply structures should be constituted. The analysis of possibilities about biorefinery location depends on transport distances and complexity of the supply system. The concern is to choose the most appropriate biomass based on types and sources, and to plan the logistic routine and network structure [34]. The concept called "integrated supply logistics"—focused on biomass production and supply systems—plays an important role in the suitable biomass use.

7.7.1 Requirements in the Biomass Supply

The biomass supply considers its production, processing, and conditioning in order to generate a material with defined properties and a strong connection between the producer and the consumer. These systems present special characteristics with respect to other bio-economy subsections, such as (i) considerable number of sources plus small supply volumes, (ii) periodicity of biomass market, and (iii) changeability in physicochemical features of biomass. The design of biomass supply systems can be proposed with different targets. These targets can be connected between them or exist in an independent way. The consumer of biomass should define the main requirements of the supply system, for example, ensuring the biomass availability along the year, having a continuous biomass flow, and choosing a suitable biomass for determined products, among others.

Ensuring the biomass availability along the year is a decisive factor in biorefinery advisability. When the supply systems are insecure and inadequate, the biorefinery viability enters into a critical state. Aspects such as having a continuous biomass flow and choosing a suitable biomass depend strongly on its sustainable availability and prices evolution in the long term, as well as periodicity of biomass market. For the biomass that depends on seasonality, a suitable storage must be taken into account. In general terms, in the strategic planning of biorefineries, the conservation of production areas of biomass must be considered.

The continuous flows of biomass in the biorefinery promote cost reduction in operation and ensure a homogeneous use of capacities for processing and logistics [33]. The continuity can be reached by using high available biomass or through the option of stock keeping of biomass. The stock keeping can be made by producers, processors, and conditioners of biomass in supply systems in order to allow continuous deliveries to consumers [33]. Particle size, geometry, calorific value, water, and ash content, among others, are relevant attributes to choose a suitable biomass due to its influence in the cohesion, performance, and profitability of the biorefinery. Additionally, the uniformity of used biomass affects the technical stability of biorefinery. These attributes significantly change for some types of biomass making homogeneity an important factor in supply systems, where concepts such as standardization, appropriate supply, and mixture of feedstock in the storage area gain attraction.

7.7.2 Supply Options

Supply options concern about adjustable parameters in a supply system. According to a certain processing technology and/or biorefinery location, options consider the types of biomass, the supply systems, and transport grid [33–37].

- Biomass portfolio used in biorefineries includes from first-generation feedstocks through fourth-generation feedstocks (see Chapter 2). The selection of feedstock depends on the biorefinery targets, focused on the products that will be obtained. Likewise, regional availability, biomass production costs, storage, and transportation are key factors in its selection. The biomass production costs include agriculture and harvesting costs.
- Biomass prices fluctuate considerably according to biomass sources and their availability. In agreement with market situation and supply method, biomass consumers can buy biomass in international field or guarantee the regional biomass supply. The continuous availability of biomass sources can be carried out through the implementation of crop rotation or through the association of biomass producers in order to promote strategic cooperation models.
- Supply systems contemplate sequences of processes for the production, conditioning, and consumption of biomass. Between these processes logistical steps such as loading, transport, unloading, and storage are included. These steps aim to create a bridge between production and consumption of biomass referred to distance and time [33]. The logistic aspect represents a considerable percentage in biomass supply costs; therefore, the searching of solutions provides the access to prudent biomass costs.
- The selection of transport method relies on biomass type, plant location, and infrastructural conditions.

7.7.3 Outlining of Supply Systems

Strategic outlining of supply systems can be defined as a set of decision problems that involves logistic factors. These problems consider baseline situations, activity alternatives, general objective of strategic outlining, and the results [35–37]. The baseline situation defines the basic setting, which cannot be affected by the planner. In a biomass supply system the following are the considerations: (i) biomass consumption area (sink), (ii) location of biomass (sources), (iii) competition for production areas and resources, (iv) biomass types to be used and its value in the biorefinery, and (v) transport grid considering land, maritime, and air transport [33].

On the contrary, to fixed baseline situation, activity alternatives are variables and are chosen. To constitute a biomass supply system, a transport system is established. The activity alternatives contain the following: (i) biomass sources, (ii) transport of goods, (iii) transport and management technique, (iv) scheme of transport grid, and (v) transport of total biomass [34].

Fixed and variable elements are related between them. This adaptation also involves the main objective of the strategic outlining that can consider one or more objectives. Examples of optimization aims are minimizing supply costs, minimizing transport costs, minimizing environmental impacts, etc. [34–[37].

7.8 Conclusion

Some biorefineries were described in this chapter. It is clear that the biorefineries existing today were boosted mainly by bioenergy demands but any way based on the existing facilities to produce food, feed, and beverage products mainly. A breakdown in the future can be to change from low-value-added, energy-driven biorefineries to high-value-added, product-driven biorefineries. However, in both cases the biomass supply should be a critical point, more than the technologies.

References

1. J. Moncada, M. M. El-Halwagi, and C. A. Cardona, "Techno-economic analysis for a sugarcane biorefinery: Colombian case," *Bioresour. Technol.*, vol. 135, pp. 533–543, 2013.
2. J. Moncada, L. G. Matallana, and C. A. Cardona, "Selection of process pathways for biorefinery design using optimization tools: A colombian case for conversion of sugarcane bagasse to ethanol, poly-3-hydroxybutyrate (PHB), and energy," *Ind. Eng. Chem. Res.*, vol. 52, no. 11, pp. 4132–4145, 2013.
3. J. A. Quintero, M. I. Montoya, O. J. Sánchez, O. H. Giraldo, and C. A. Cardona, "Fuel ethanol production from sugarcane and corn: Comparative analysis for a Colombian case," *Energy*, vol. 33, no. 3, pp. 385–399, 2008.
4. E. C. Rivera, A. C. Costa, D. I. P. Atala, F. Maugeri, M. R. W. Maciel, and R. M. Filho, "Evaluation of optimization techniques for parameter estimation: Application to ethanol fermentation considering the effect of temperature," *Process Biochem.*, vol. 41, pp. 1682–1687, Jul. 2006.

5. M. Balat, M. Balat, E. Kırtay, and H. Balat, "Main routes for the thermo-conversion of biomass into fuels and chemicals. Part 2: Gasification systems," *Energy Convers. Manag.*, vol. 50, pp. 3158–3168, Dec. 2009.
6. L. E. Rincón, L. A. Becerra, J. Moncada, and C. A. Cardona, "Techno-economic analysis of the use of fired cogeneration systems based on sugar cane bagasse in south eastern and mid-western regions of Mexico," *Waste Biomass Valori.*, vol. 5, no. 2, pp. 189–198, 2014.
7. S. I. Mussatto, J. Moncada, I. C. Roberto, and C. A. Cardona, "Techno-economic analysis for brewer's spent grains use on a biorefinery concept: The Brazilian case," *Bioresour. Technol.*, vol. 148, pp. 302–310, 2013.
8. S. Shahhosseini, "Simulation and optimisation of PHB production in fed-batch culture of *Ralstonia eutropha*," *Process Biochem.*, vol. 39, no. 8, pp. 963–969, 2004.
9. W. Zhang, M. Seki, S. Furusaki, and A. P. J. Middelberg, "Anthocyanin synthesis, growth and nutrient uptake in suspension cultures of strawberry cells," *J. Ferment. Bioeng.*, vol. 86, no. 1, pp. 72–78, 1998.
10. K. Wang, L. Ou, T. Brown, and R. C. Brown, "Beyond ethanol: A techno-economic analysis of an integrated corn biorefinery for the production," *Biofuels, Bioprod. biorefining*, vol. 9, pp. 190–200, 2015.
11. J. Moncada, J. Tamayo, and C. A. Cardona, "Evolution from biofuels to integrated biorefineries: Techno-economic and environmental assessment of oil palm in Colombia," *J. Clean. Prod.*, vol. 81, pp. 51–59, 2014.
12. V. Aristizábal, C. A. García, and C. A. Cardona, "Integrated production of different types of bioenergy from oil palm through biorefinery concept," *Waste Biomass Valori.*, vol. 7, no. 4, pp. 737–745, 2016.
13. N. Leksawasdi, E. L. Joachimsthal, and P. L. Rogers, "Mathematical modelling of ethanol production from glucose/xylose mixtures by recombinant *Zymomonas mobilis*," *Biotechnol. Lett.*, vol. 23, pp. 1087–1093, 2001.
14. L. E. Rincón, V. Hernández, and C. A. Cardona, "Analysis of technological schemes for the efficient production of added-value products from Colombian oleochemical feedstocks," *Process Biochem.*, vol. 49, pp. 474–489, Mar. 2014.
15. C. D. Botero, D. L. Restrepo, and C. A. Cardona, "A comprehensive review on the implementation of the biorefinery concept in biodiesel production plants," *Biofuel Res. J.*, vol. 4, no. 3, pp. 691–703, Sep. 2017.
16. "Biodiesel Fill Stations and Prices." Available at http://www.usabiodieselprices.com/station_map.php. Accesed April 22, 2018.
17. "Sigma-Aldrich." Available at https://www.sigmaaldrich.com. Accesed April 22, 2018.
18. Nasdaq. GlobeNewswire, "Hongli clean energy technologies corp. reports fiscal year 2016 Q2 financial results Nasdaq: CETC."
19. "Fisher Scientific: Lab equipment and supplies." Available at https://www.fishersci.com/us/en/home.html. Accessed April 3, 2018.
20. "EPM | Empresa de servicios públicos de Colombia." Available at https://www.epm.com.co/site/. Accessed April 9, 2018.
21. G. W. Huber, J. N. Chheda, C. J. Barrett, and J. A. Dumesic, "Production of liquid alkanes by aqueous-phase processing of biomass-derived carbohydrates," *Science*, vol. 308, pp. 1446–1450, 2005.
22. V. Aristizábal, Á. Gómez, and C. A. Cardona, "Biorefineries based on coffee cut-stems and sugarcane bagasse: Furan-based compounds and alkanes as interesting products," *Bioresour. Technol.*, vol. 196, pp. 480–489, 2015.

23. J. Moncada, J. A. Tamayo, and C. A. Cardona, "Integrating first, second, and third generation biorefineries: Incorporating microalgae into the sugarcane biorefinery," *Chem. Eng. Sci.*, vol. 118, pp. 126–140, 2014.
24. J. A. Quintero, J. Moncada, and C. A. Cardona, "Techno-economic analysis of bioethanol production from lignocellulosic residues in Colombia: A process simulation approach," *Bioresour. Technol.*, vol. 139, pp. 300–307, Jul. 2013.
25. J. C. Carvajal, Á. Gómez, and C. A. Cardona, "Comparison of lignin extraction processes: Economic and environmental assessment," *Bioresour. Technol.*, vol. 214, pp. 468–476, 2016.
26. S. Laurichesse and L. Avérous, "Chemical modification of lignins: Towards biobased polymers," *Prog. Polym. Sci.*, vol. 39, pp. 1266–1290, 2014.
27. J. Moncada Botero, "Design and evaluation of sustainable biorefineries from feedstocks in tropical regions," Master thesis. Departamento de Ingeniería Química, Universidad Nacional de Colombia sede Manizales, 2012.
28. F. Cherubini, "The biorefinery concept: Using biomass instead of oil for producing energy and chemicals," *Energy Convers. Manag.*, vol. 51, no. 7, pp. 1412–1421, 2010.
29. A. A. González, J. Moncada, A. Idarraga, M. Rosenberg, and C. A. Cardona, "Potential of the Amazonian exotic fruit for biorefineries: The *Theobroma bicolor* (Makambo) case," *Ind. Crops Prod.*, vol. 86, pp. 58–67, 2016.
30. Y. Chisti, "Biodiesel from microalgae beats bioethanol," *Trends Biotechnol.*, vol. 26, no. 3, pp. 126–131, 2008.
31. R. Maria, D. B. Alves, C. Augusto, J. F. O. Granjo, B. P. D. Duarte, and M. C. Oliveira, "Kinetic models for the homogeneous alkaline and acid catalysis in biodiesel production," *10th Int. Symp. Process Syst. Eng.—PSE2009*, vol. 27, pp. 483–488, 2009.
32. L. E. Rincón, J. J. Jaramillo, and C. A. Cardona, "Comparison of feedstocks and technologies for biodiesel production: An environmental and techno-economic evaluation," *Renew. Energy*, vol. 69, pp. 479–487, 2014.
33. P. Fiedler, M. Lange, and M. Schultze, "Supply logistics for the industrialized use of biomass: Principles and planning approach," *LINDI 2007—Int. Symp. Logist. Ind. Informatics 2007, Proc.*, vol. 1, pp. 41–46, 2007.
34. A. E. Duarte, W. A. Sarache, and Y. J. Costa, "A facility-location model for biofuel plants: Applications in the Colombian context," *Energy*, vol. 72, pp. 476–483, 2014.
35. D. Yue, F. You, and S. W. Snyder, "Biomass-to-bioenergy and biofuel supply chain optimization: Overview, key issues and challenges," *Comput. Chem. Eng.*, vol. 66, pp. 36–56, 2014.
36. B. Sharma, R. G. Ingalls, C. L. Jones, and A. Khanchi, "Biomass supply chain design and analysis: Basis, overview, modeling, challenges, and future," *Renew. Sustain. Energy Rev.*, vol. 24, pp. 608–627, 2013.
37. S. D. Ekşioĝlu, A. Acharya, L. E. Leightley, and S. Arora, "Analyzing the design and management of biomass-to-biorefinery supply chain," *Comput. Ind. Eng.*, vol. 57, no. 4, pp. 1342–1352, 2009.

8

Sustainability Assessment of Biorefineries Based on Indices

The indices for biorefineries are considered determinants to define the most promising configurations and to judge their technological and economic risks. These indices reflect one principle of sustainability: integral use of the renewable raw materials. A simple calculation can be performed in order to account for how efficient is a biorefinery system or process in terms of the conversion of biomass into final products. In this way, this chapter presents the indices required to assess a biorefinery considering its complexity, conversion capacity, water and reagent consumption, and economic results. For the technical assessment of a biorefinery, indices such as biorefinery mass index, biorefinery water mass index, biorefinery reagents mass index, biorefinery coproducts and by-products mass index, biorefinery energy index, and biorefinery energy index by equipment were proposed. In addition, for the economic assessment of a biorefinery, the index called biorefinery economic index was proposed. All the abovementioned indices are explained in detail, and some case studies are presented to show how easy the calculations are.

8.1 CCS Biorefinery

Refinery size is a function of distillation capacity. Relative size can be a function of refinery complexity [1]. In the 1960s, W. L. Nelson developed the complexity index, which allows us to quantify the relative cost of refinery components with respect to its conversion capacity. Nelson's index compares the costs of evolutionary units (e.g., catalytic cracking unit or catalytic reformer) with the costs of primary distillation [1]. The index determination is an effort to calculate the relative cost of a refinery as a function of additional cost of evolutionary units and relative processing capacity. The value of Nelson's complexity index is proportional to the relative cost of the refinery. Analogous to this, in this chapter an index called "biorefinery complexity index (BCI)" is assessed. BCI is calculated based on platforms, feedstocks, products, and processes considered in a biorefinery, and it is useful to determine the most promising configurations and to judge the technological and economic risks [2].

A biorefinery is composed mainly of features such as platforms, feedstocks, products, and processes [2]. To determine the technical and economic state of a biorefinery using the abovementioned features, a "feature complexity index (FCI)" is developed. The FCI is determined based on the evaluation of the "technology readiness level (TRL)" for each feature of the biorefinery. In the case of "products" and "feedstocks" that can be or will be commodities in the market, these are evaluated according to the "market readiness level (MRL)" that is analogous to TRL. The combination of FCI for platforms, feedstocks, products, and processes gives the BCI.

Therefore, calculation of the BCI is explained later. In each of the four features of a biorefinery, the TRL is assessed taking into account a description level in a range of 1 (basic research) to 9 (system proven and ready for full commercial deployment) [2]. Considering the TRL, the feature complexity (FC) for each feature of the biorefinery is determined by Equation 8.1. The FCI for platforms, feedstocks, products, and processes is calculated using the number of features (NF) and FC of each one (Equation 8.2). The BCI is the sum of the four FCIs as indicated in Equation 8.3. Finally, the "biorefinery complexity profile (BCP)" is given by the BCI and the four FCIs as show in Equation 8.4.

$$FC_i = 10 - TRL_i \tag{8.1}$$

$$FCI_i = \sum_{j=1}^{m} NF_i * FC_i \tag{8.2}$$

$$BCI = \sum_{j=1}^{m} FCI_i \tag{8.3}$$

$$BCP = BCI \left(FCI_{feedstocks} / FCI_{platforms} / FCI_{processes} / FCI_{products} \right) \tag{8.4}$$

8.2 Mass and Energy Biorefinery Indices

8.2.1 Biorefinery Mass Indices

This index relates the ratio of the mass of valuable products to fresh biomass fed into the biorefinery. This index is called biorefinery mass index (MI_B), and it is defined by Equation 8.5. An MI_B is good when the value tends to 1, namely the yields are high.

Other alternative indices for MI_B are biorefinery water mass index $\left(MI_B^w \right)$, biorefinery reagents mass index $\left(MI_B^r \right)$, and biorefinery coproducts and by-products mass index $\left(MI_B^{cop-byp} \right)$. MI_B^w considers the ratio of the mass of valuable products to water input streams and biomass fed. MI_B^r determines the ratio of the mass of valuable products to biomass fed and chemical reagent inputs (i.e., sulfuric acid, sodium hydroxide, etc.). $MI_B^{cop-byp}$ relates the mass of valuable coproducts and by-products to fresh biomass fed. The goal of these indices is to assess the technical efficiency of a biorefinery in detail and show the effect with the addition of other streams to the calculation. Then, MI_B^w, MI_B^r, and $MI_B^{cop-byp}$ are defined by Equations 8.6, 8.7, and 8.8, respectively.

$$MI_B = \frac{\sum_{i=1}^{n} m_i^p}{\sum_{j=1}^{n} m_j^f} \tag{8.5}$$

$$MI_B^w = \frac{\sum_{i=1}^{n} m_i^p}{\sum_{j=1}^{n} m_j^w + \sum_{j=1}^{n} m_j^f} \tag{8.6}$$

$$MI_B^r = \frac{\sum_{i=1}^{n} m_i^p}{\sum_{j=1}^{n} m_j^r + \sum_{j=1}^{n} m_j^f} \tag{8.7}$$

$$MI_B^{cop-byp} = \frac{\sum_{i=1}^{n} m_i^{cop} + \sum_{i=1}^{n} m_i^{byp}}{\sum_{j=1}^{n} m_j^f} \tag{8.8}$$

where:

- i denotes the species i, referring to products, coproducts, and by-products.
- j denotes the species j, referring to feedstock, water, and reagents.
- m denotes the mass flow rate of feedstock, water, reagents, products, coproducts, and by-products; superscripts f, w, r, p, cop, and byp denote feedstock, water, reagents, products, co-products, and by-products, respectively.

8.2.2 Biorefinery Energy Indices

For this index, it is possible to consider two calculations taking into account the energy required in the biorefinery based on the raw material and products. The first one relates the gross energy potential of valuable products to fresh biomass fed. The gross energy potential (E) is defined by Equation 8.9, and it is the energy stored in a raw material or a product in its primary form and can be used in a transformation process. This index can be called biorefinery energy index (EnI_B) and is defined by Equation 8.10. This index reflects other principles of sustainability from the energy point of view and will be favorable, for example, when the production of biofuels is considered (i.e., ethanol, butanol, biodiesel, biogas, hydrogen, etc.), producing more energy than that of the initial raw material. The second one relates the energy required per equipment to fresh biomass fed. This index is called the biorefinery energy index by equipment ($EnI_{B-equip}$) and is defined by Equation 8.11.

$$E = m*HHV \tag{8.9}$$

$$EnI_B = \frac{\sum_{i=1}^{n} E_i^p}{\sum_{j=1}^{n} E_j^f} \tag{8.10}$$

$$EnI_{B-equip} = \frac{\sum_{j=1}^{n} En_j^{equip}}{\sum_{j=1}^{n} m_j^f} \tag{8.11}$$

where:

E denotes the gross energy potential.

HHV denotes the higher heating value (MJ/kg).

En denotes the heat required by the equipment.

8.3 Economic Biorefinery Indices

This index relates the production cost with the sale price of valuable products in the biorefinery or stand-alone process. This index can be called biorefinery economic index (EcI_B), and it is defined by Equation 8.12. An EcI_B is good when the value tends to zero, namely the gains are positive. When the coproducts and by-products are considered, Equation 8.12 is transformed into Equation 8.13.

$$EcI_{B1} = \frac{\sum_{i=1}^{n} pc_i^p}{\sum_{j=1}^{n} sp_j^p} \tag{8.12}$$

$$EcI_{B2} = \frac{\sum_{i=1}^{n} pc_i^p + \sum_{i=1}^{n} pc_i^{cop} + \sum_{i=1}^{n} pc_i^{byp}}{\sum_{j=1}^{n} sp_j^p + \sum_{j=1}^{n} sp_j^{cop} + \sum_{j=1}^{n} sp_j^{byp}} \qquad (8.13)$$

where:

i,j denote the species i, referring to products, coproducts, and by-products.

pc denotes the production cost of products, coproducts, and byproducts; and the superscripts p and byp denote products and by-products, respectively.

sp denotes the sale price of products, coproducts, and byproducts; and superscripts p and byp denote products and by-products, respectively.

8.4 Case Studies

In order to understand the concepts explained earlier, some study cases are presented below. The indices are calculated for three biorefineries that consider sugarcane, oil palm fresh fruit brunches, and coffee cut-stems (CCSs) as feedstocks. The biorefinery from sugarcane and CCSs includes an analysis of economic indices when there are different scales of processing of feedstock.

8.4.1 Sugarcane Biorefinery

Three biorefineries utilizing sugarcane for obtaining sugar, ethanol, electricity, xylitol, and polyhydroxybutyrate (PHB) are depicted in Figures 8.1–8.3. The constituent plants of the biorefinery as a base case (Scenario 1) are aimed at obtaining sugar, ethanol, and electricity. While, Scenarios 2 and 3 include xylitol production, and xylitol and PHB production, respectively, to the base case. For these scenarios, a plant for the production of sugars (xylose and glucose) is required for obtaining the platforms for both xylitol and PHB plants, respectively. A cogeneration plant was considered for supplying part of the energy requirements of the biorefinery. The inlet flow of sugarcane was set at 10,000 kg/h, and the biorefineries were assumed to operate 8,000 h/year. The purpose, operating conditions, assumptions, and methods used for the principal constituent units in the simulations are summarized in Table 8.1.

The authors present the calculations of BCI and BCP for Scenario 3 because it is the biggest sugarcane biorefinery in comparison with Scenarios 1 and 2.

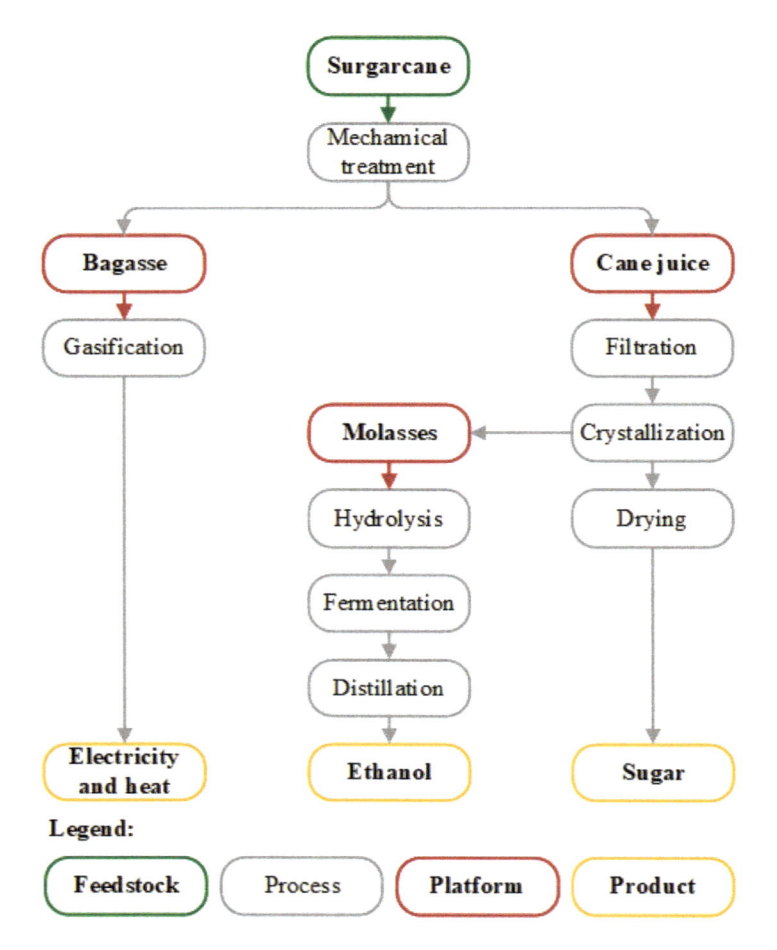

FIGURE 8.1
Flowsheet of sugarcane biorefinery, Scenario 1.

Figure 8.3 shows the features of biorefinery from sugarcane (Scenario 3) in terms of platforms, feedstocks, products, and processes. Jungmeier et al. (2014) indicates the TLR values for each single feature, and these values depend on the specifications of the feature [2]. According to the features of sugarcane biorefinery, TRL values are

- 1 feedstock with $TRL_{feedstock}$ of 9 ($NF_{feedstock} = 1$).
- 7 platforms with $TRL_{platforms}$ of 9, 9, 9, 7, 8, 7 and 9 ($NF_{platform} = 7$).
- 17 processes with $TRL_{processes}$ of 9, 8, 8, 8, 7, 9, 8, 8, 9, 8, 8, 8, 9, 8, 8, 9 and 9 ($NF_{process} = 17$).
- 5 products with $TRL_{products}$ of 9, 8, 9, 9 and 9 ($NF_{products} = 5$).

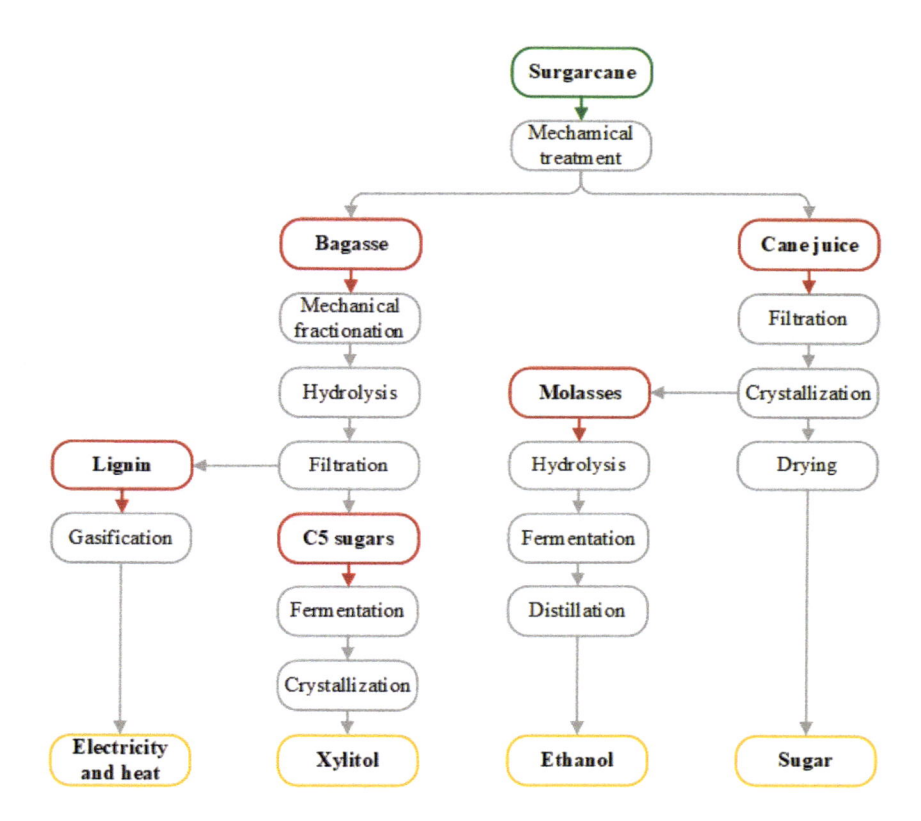

FIGURE 8.2
Flowsheet of sugarcane biorefinery, Scenario 2.

Considering Equation 8.1, the feature complexities are as follows:

- $FC_{feedstock} = (10 - 9) = 1.$
- $FC_{platform1} = (10 - 9) = 1$, $FC_{platform2} = (10 - 9) = 1$, $FC_{platform3} = (10 - 9) = 1$, $FC_{platform4} = (10 - 9) = 1$, $FC_{platform5} = (10 - 8) = 2$, $FC_{platform6} = (10 - 7) = 3$ and $FC_{platform7} = (10 - 9) = 1.$
- $FC_{process1} = (10 - 9) = 1$, $FC_{process2} = (10 - 8) = 2$, $FC_{process3} = (10 - 8) = 2$, $FC_{process4} = (10 - 8) = 2$, $FC_{process5} = (10 - 7) = 3$, $FC_{process6} = (10 - 9) = 1$, $FC_{process7} = (10 - 8) = 2$, $FC_{process8} = (10 - 8) = 2$, $FC_{process9} = (10 - 9) = 1$, $FC_{process10} = (10 - 8) = 2$ and $FC_{process11} = (10 - 8) = 2$, $FC_{process12} = (10 - 8) = 2$, $FC_{process13} = (10 - 9) = 1$, $FC_{process14} = (10 - 8) = 2$, $FC_{process15} = (10 - 8) = 2$, $FC_{process16} = (10 - 9) = 1$ and $FC_{process17} = (10 - 9) = 1.$
- $FC_{product1} = (10 - 9) = 1$, $FC_{product2} = (10 - 8) = 2$, $FC_{product3} = (10 - 9) = 1$, $FC_{product4} = (10 - 9) = 1$ and $FC_{product5} = (10 - 9) = 1.$

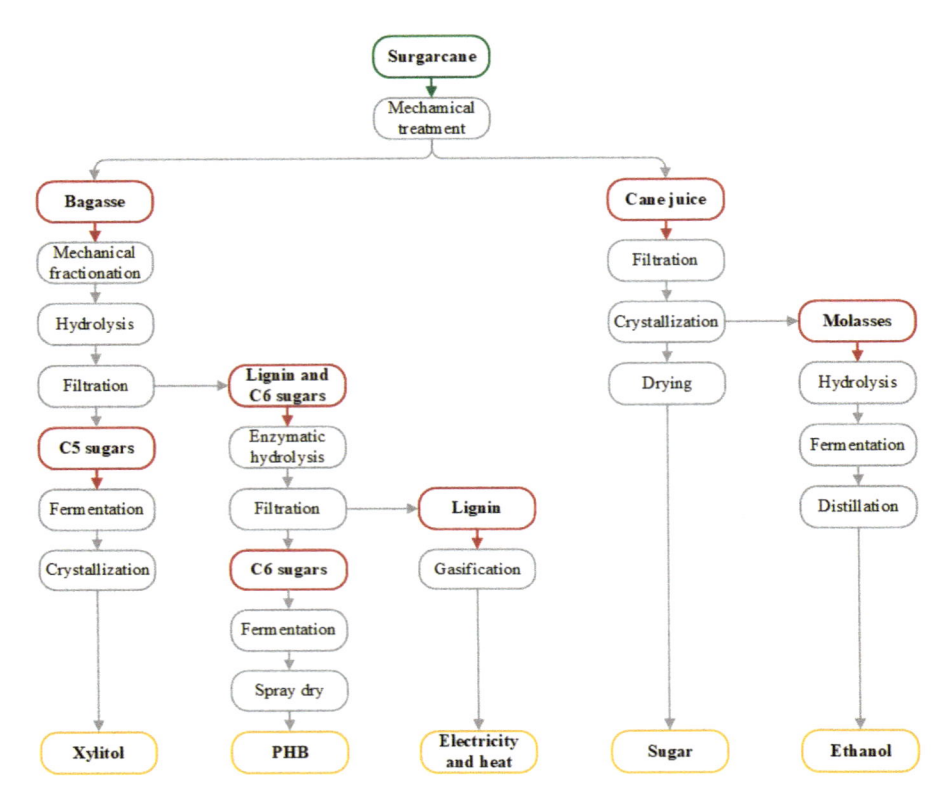

FIGURE 8.3
Flowsheet of sugarcane biorefinery, Scenario 3.

Considering Equation 8.2, the FCIs for the four features are as follows:

- $FCI_{feedstock} = (1*1) = 1$.
- $FCI_{platforms} = (1*1) + (1*1) + (1*1) + (1*3) + (1*2) + (1*3) + (1*1) = 12$.
- $FCI_{processes} = (1*1) + (1*2) + (1*2) + (1*2) + (1*3) + (1*1) + (1*2) + (1*2) + (1*1)$
  $+ (1*2) + (1*2) + (1*2) + (1*1) + (1*2) + (1*2) + (1*1)+(1*1) = 29$.
- $FCI_{products} = (1*1) + (1*2) + (1*1) + (1*1) + (1*1) = 6$.

Considering Equation 8.3, the BCI is

$$BCI = 1 + 12 + 29 + 6 = 48$$

Finally, according to Equation 8.4, the BCP of the biorefinery from sugarcane in Scenario 3 is 48(1/12/29/6). For Scenarios 1 and 2, the BCP is 19(1/3/12/3) and 36(1/9/22/4), respectively. As can be seen, the addition of processes (xylitol and PHB) to the base scenario influences the BCP positively, increasing

TABLE 8.1

Purpose and Operating Conditions Used for the Main Units in the Simulation of the Sugarcane Biorefinery

Unit	Purpose	Operating Conditions	Assumptions	Reference
Sugar Milling				
Milling	Separating bagasse and molasses	The sucrose extraction yield reached 95%.	N.A.	[3]
Clarifier	Removing impurities	110°C and SO_2	NRTL-HOC	
Evaporator	Concentration of juice	Removing up to 75% of the water	NRTL-HOC	
Crystallizer	Formation of sugar crystals	25°C, 1 bar	NRTL	
Sugar Plant				
Milling	Size reduction to 0.45 mm	1 bar, Jaw mill	N.A.	
Acid hydrolysis	Recovering xylose contained in bagasse	121°C, 1 bar, 2% v/v of H_2SO_4, 1:10 solid–liquid ratio	User model (Yields from literature)	[4]
Detoxification	Neutralizing the inhibitors produced in acid hydrolysis (i.e., HMF, furfural)	25°C, 1 bar, pH 6.5 $Ca(OH)_2 + H_2SO_4 \rightarrow CaSO_4 + 2H_2O$	User model (yields from the literature)	[5]
Enzymatic hydrolysis	Recovering glucose contained in bagasse	50°C, 1 bar, agitated tank using endo β-1,4,glucanases	User model (yields from the literature)	[6]
Ethanol Plant				
Fermentation	Ethanol production	37°C using *Saccharomyces cerevisiae*	User model (yields from the literature)	[7]
Distillation columns	Ethanol separation	*Distillation.* 12 trays, 6 feed stray, 1.8 reflux ratio, total condenser and 1 bar. *Rectification.* 16 trays, 8 feed stray, 1.8 reflux ratio, total condenser and 1 bar.	NRTL-HOC	[8]
Cogeneration Plant				
Gasifier	Syngas generation	500°C, 60 bar and a ratio of 1.1125 air-to-fuel	NRTL-HOC	[9]
Turbine	Electricity generation	1 bar, 80% efficiency	NRTL-HOC	[10]
Xylitol Plant				
Fermentation	Xylitol production	30°C using *Candida mogii*	User model (yields from the literature)	[11]

(Continued)

TABLE 8.1 (*Continued*)

Purpose and Operating Conditions Used for the Main Units in the Simulation of the Sugarcane Biorefinery

Unit	Purpose	Operating Conditions	Assumptions	Reference
Crystallizer	Xylitol purification	–5°C, 1 bar	NRTL	[12]
PHB Plant				
Fermentation	PHB production	30°C using *Ralstonia eutropha*	User model (yields from the literature)	[13]
Spray drying	PHB purification	50°C, 1 bar	NRTL	[14]

NRTL: Non-random Two-Liquid model for activity coefficients calculation.
HOC: Hayden-O'Conell equation of state used for description of the vapor phase.
N.A: Not applicable.

their values, mainly the number of platforms and processes. It means that Scenarios 2 and 3 have higher conversion capacity and are more complex than Scenario 1. The BCI is useful to determine the most promising configuration and to judge the technological and economic risks. Figure 8.4 indicates the BCI and BCP of sugarcane biorefineries taking into account the number of feedstocks, platforms, processes, and processes that each system considers.

Figure 8.5 shows the mass indices for three scenarios of the sugarcane biorefinery. The MI_B presents an upward trend, as the products are added in each scenario this index increases in a logic way, improving the global yield of biorefinery. In the base scenario, the individual yields for sugar and ethanol are 0.11 kg of sugar per kilogram of sugarcane and 62.4 L of ethanol per ton of sugarcane, respectively. The yields obtained in simulation are close to the values reported in the literature, 0.12 kg of sugar per kilogram of sugarcane [15] and 70 l of ethanol per ton of sugarcane [16].

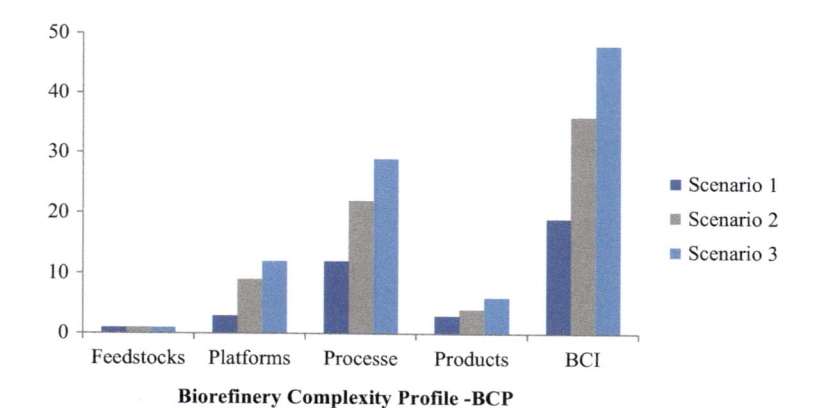

Biorefinery Complexity Profile -BCP

FIGURE 8.4
Biorefinery complexity profiles of sugarcane biorefineries.

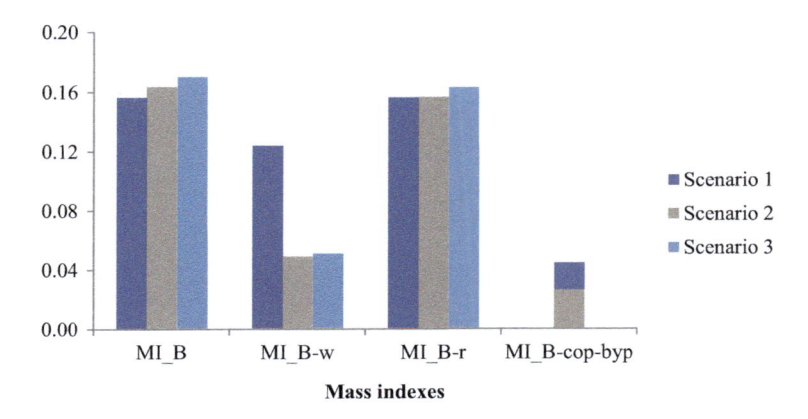

FIGURE 8.5
Mass indices for sugarcane biorefineries.

In biotechnological processes, water consumption is presented as a critical point due to high requirements. In this case, the MI_B^w shows low values for all scenarios indicating that the water consumption is considerable. The addition of products in Scenarios 2 and 3 affects the water index, reporting lower values than the base scenario. Scenarios 2 and 3 consider the pretreatment stage of sugarcane bagasse, which demands large volumes of liquid in order to extract sugars from lignocellulosic materials that are the substrate to obtain xylitol and PHB. For the case of MI_B^r, the ratio between the products and reagent demand and raw material is very similar for all scenarios. The obtained values indicate that the demand of reagents is higher than the flow of the product.

In all scenarios, the generated electricity is taken as a coproduct, and as can be seen when the sugarcane bagasse is only dedicated to electricity generation (Scenario 1), the $MI_B^{cop-byp}$ is the highest. This index is calculated with the electricity that remains after supplying the requirements of the processes considered in each biorefinery, namely, the electricity that can be sold. For Scenario 3, the remaining electricity is minimum, for this reason the index is near to zero.

For Scenarios 1, 2, and 3, the $EnI_{B-equip}$ is 8.16, 13.24, and 15.76, respectively. The amount of energy required for each scenario is directly proportional to the addition of products. Namely, the energy requirements of the equipment increase with the number of products. As can be seen from the values obtained for Scenarios 2 and 3, the addition of pretreatment stage of sugarcane bagasse, xylitol, and PHB production contribute significantly to the energy demand of biorefineries, affecting negatively the energy index. However, the cogeneration system considered in the biorefineries mitigates these requirements with the production of low- and medium-pressure steam.

After obtaining the mass and energy indices, an analysis of the EcI_B as a function of processing capacity for all scenarios is done in order to see the trend and define the economic performance of the same biorefineries at small scale.

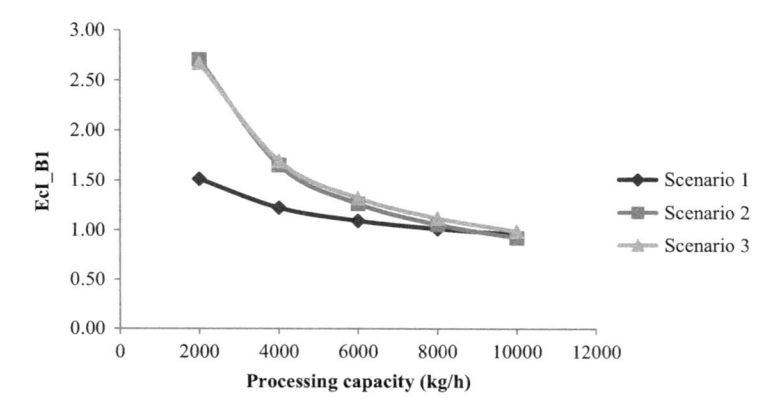

FIGURE 8.6
Economic index versus processing capacity for SC biorefineries.

Figure 8.6 indicates the results obtained in this analysis, EcI_B versus processing capacity for all SC scenarios. The decrease in processing capacity affects the economic index of all scenarios in a negative way. The addition of xylitol and PHB to Scenarios 2 and 3 increases the economic index, making the systems unfeasible economically. However, if the biorefineries consider processing capacities higher than 10 ton/h, then there exists the possibility to obtain better economic results. These results allow us to conclude that the biorefinery is not feasible, from an economic point of view, if it is carried out at small scale.

Two more case studies are discussed later, where oil palm and CCSs are used as feedstocks. Initially, the biorefineries are described, and then the obtained results are indicated. Because BCI and BCP are comparative indices and the proposed biorefineries consider a single configuration, the analysis of these results is based on the comparison of features of both biorefineries.

8.4.2 Oil Palm Biorefinery

The biorefinery from fresh fruit bunches (FFBs) considers the production of biodiesel, ethanol, butanol, and hydrogen and obtains acetone as a coproduct and electricity as a by-product [17]. Figure 8.7 indicates the flowsheet of biorefinery proposed and as can be seen, the biorefinery is composed of 1 feedstock, 4 platforms, 11 processes, and 6 products. The feed of raw material is 1 ton/h. The biorefinery is composed of seven processes. Initially, FFBs enter the extraction stage where the palm oil and cake are obtained. The palm oil is subjected to pre-esterification and transesterification reactions for the biodiesel production, and the glycerol obtained as a by-product in this stage is the substrate to obtain ethanol through fermentation using *Escherichia coli* as a microorganism. The cake is rich in lignocellulosic material and is sent to the pretreatment stage where particle size reduction and enzymatic hydrolysis for the extraction of sugars are carried out. A hexose-rich liquor (50%) is

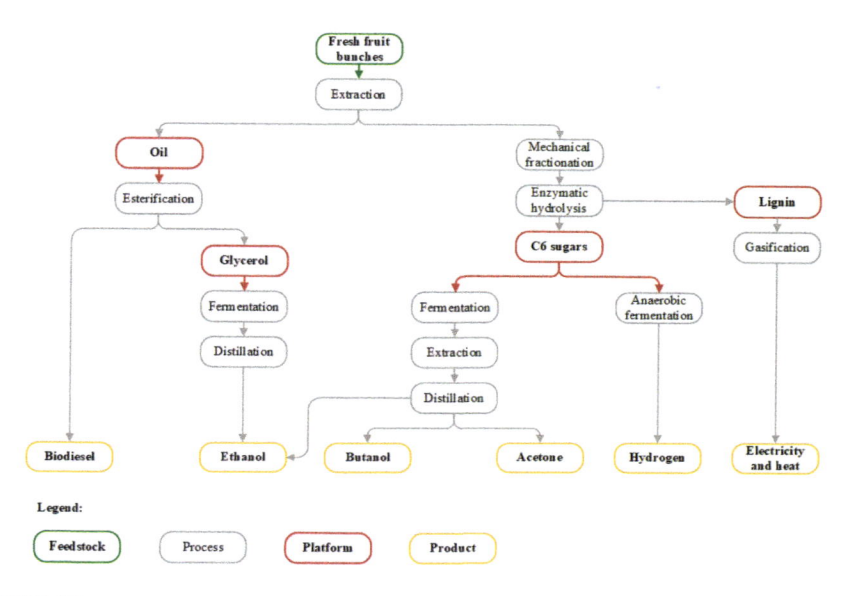

FIGURE 8.7
Flowsheet of FFB biorefinery.

used as a substrate to obtain butanol by acetone, butanol and ethanol (ABE) fermentation using *Clostridium acetobutylicum*. The remaining hexose-rich liquor is the raw material to obtain hydrogen via dark fermentation using an anaerobic digestion model. The solid material obtained in the pretreatment stage is used in the generation of electricity by gasification. Table 8.2 shows the mass and energy balances of the biorefinery from FFB.

TABLE 8.2

Information of Energy and Mass Balance of FFB Biorefinery

Biorefinery	FFB
Raw material	FFB: 1000 kg/h, HHV: 16.7 MJ/kg
Reagents (kg/h)	Water: 15096.6, H_2SO_4: 0.153, NaOH: 3.767, Methanol: 28.5, Dodecanol: 200, *E. coli*: 0.213, *C. acetobutylicum*: 0.102, Biomass: 1.15
Outputs (kg/h)	Water: 570.002, Waste: 15051.9, CO_2: 17.67, $CaSO_4$: 1249.21, *E. coli*: 1.47, *C. acetobutylicum*: 37.33, Biomass: 6.87, Ash: 36.06, Hot gases: 285.95
Product, coproduct, by-product	Biodiesel: 222.5kg/h, HHV: 41MJ/kg Hydrogen: 5.72kg/h, HHV: 141.9MJ/kg Butanol: 9.2kg/h, HHV: 33.075MJ/kg Ethanol: 14.5kg/h, HHV: 29.7MJ/kg Acetone: 9.2, HHV: 28.548MJ/kg Electricity: 63.85kW
Energy required by equipment (MJ/h)	5623.6

8.4.3 CCS Biorefinery

The biorefinery from CCSs considers the production of ethanol, furfural, and hydroxymethylfurfural (HMF) [18]. Figure 8.8 indicates the flow-sheet of the proposed biorefinery and as can be seen, the biorefinery is composed of 1 feedstock, 2 platforms, 11 processes, and 3 products. The biorefinery is designed taking into account different scales of processing of raw materials such as, 5, 25, 50, and 100 ton/h. The biorefinery comprises four processes. Initially, the lignocellulosic biomass is subjected to a sugar extraction process divided into three stages: (i) size reduction, (ii) dilute-acid pretreatment, and (iii) enzymatic hydrolysis. Then, 50% of glucose-rich liquor is sent to fermentation using *Saccharomyces cerevisiae* as a microorganism for ethanol production, and distillation and molecular

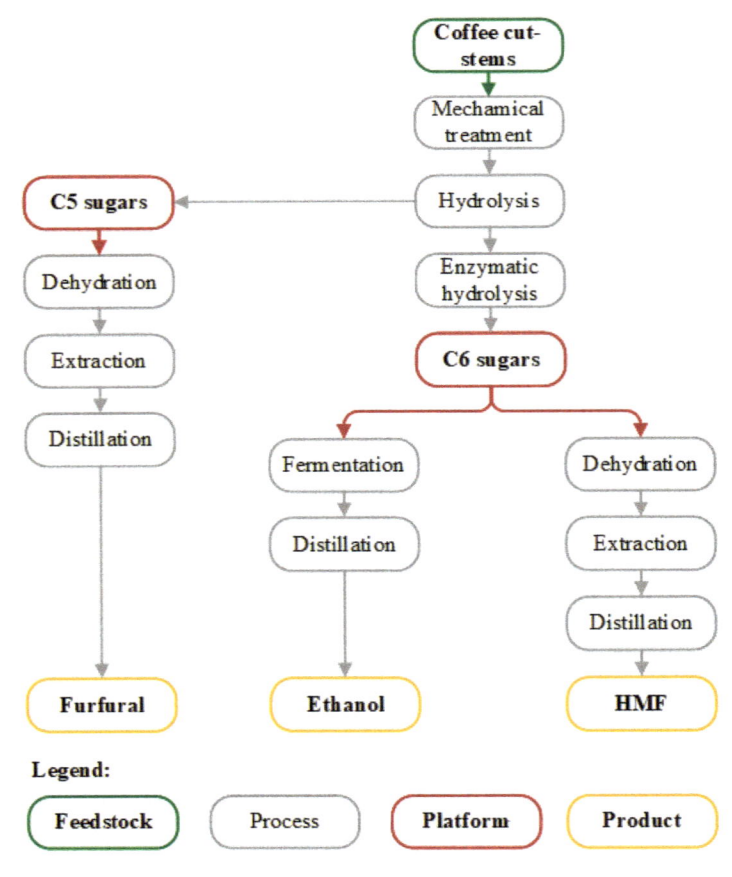

FIGURE 8.8
Flowsheet of coffee cut-stems biorefinery.

TABLE 8.3

Information of Energy and Mass Balance of CCS Biorefinery

Biorefinery	CCS (5)	CCS (25)	CCS (50)	CCS (100)
Raw material	CCS: 5,000 kg/h HHV: 19.32 MJ/kg	CCS: 25,000 kg/h HHV: 19.32 MJ/kg	CCS: 50,000 kg/h HHV: 19.32 MJ/kg	CCS: 100,000 kg/h HHV: 19.32 MJ/kg
Reagents (kg/h)	Water: 68,038.73 H_2SO_4: 900 DMSO: 550 Toluene: 930 *Saccharomyces cerevisiae:* 0.0083	Water: 340,214.2 H_2SO_4: 4,500 DMSO: 2,179.8 Toluene: 4,700 *Saccharomyces cerevisiae:* 0.0294	Water: 680,670.32 H_2SO_4: 9,000 DMSO: 5,482.67 Toluene: 9,500 *Saccharomyces cerevisiae:* 0.0819	Water: 1,360,945.51 H_2SO_4: 18000 DMSO: 10,954.44 Toluene: 20,000 *Saccharomyces cerevisiae:* 0.1635
Outputs (kg/h)	Solids: 1,582.82 Waste: 71,258.96 DMSO-traces: 636.96 CO_2: 305.50 *Saccharomyces cerevisiae:* 80.90	Solids: 6,307.397 Waste: 359,369.34 DMSO-traces: 2,008.318 CO_2: 1,224.037 *Saccharomyces cerevisiae:* 289.439	Solids: 15,845.68 Waste: 714,749.5 DMSO-traces: 5,055.52 CO_2: 3,064.513 *Saccharomyces cerevisiae:* 800.16	Solids: 31,654.33 Waste: 1,429,397.43 DMSO-traces: 10,101 CO_2: 6,113.22 *Saccharomyces cerevisiae:* 1,594.73
Product, coproduct, by-product	Furfural: 833.1 HHV: 21 MJ/kg HMF: 391.3 HHV: 22.06 MJ/kg Ethanol: 329.2 HHV: 29.7 MJ/kg	Furfural: 4,198.3 HHV: 21 MJ/kg HMF: 1,557 HHV: 22.06 MJ/kg Ethanol: 1,640.2 HHV: 29.7 MJ/kg	Furfural: 7,931.2 HHV: 21 MJ/kg HMF: 3,916.2 HHV: 22.06 MJ/kg Ethanol: 3290.3 HHV: 29.7 MJ/kg	Furfural: 16,650 HHV: 21 MJ/kg HMF: 7,824.7 HHV: 22.06 MJ/kg Ethanol: 6,564.7 HHV: 29.7MJ/kg
Energy required by equipment (MJ/h)	61,674.2	307,123.6	614,384.7	12,31332.9

sieves are used for purification. The remaining 50% of glucose is sent to HMF production by dehydration reaction, and a liquid–liquid extraction with dimethyl sulfoxide (DMSO) is used for purification. Finally, the xylose-rich liquor is sent to furfural production by dehydration reaction, and a liquid–liquid extraction with toluene is used for purification. Table 8.3 shows the mass and energy balances of the biorefinery from CCS for all processing scales of the raw material.

When Equations 8.1–8.4 are applied in the FFB and CCS biorefineries, BCPs are obtained. For FFB and CCS biorefineries, the BCPs are 34(1/6/19/8) and 27(1/4/19/3), respectively. Figure 8.9 indicates the BCI

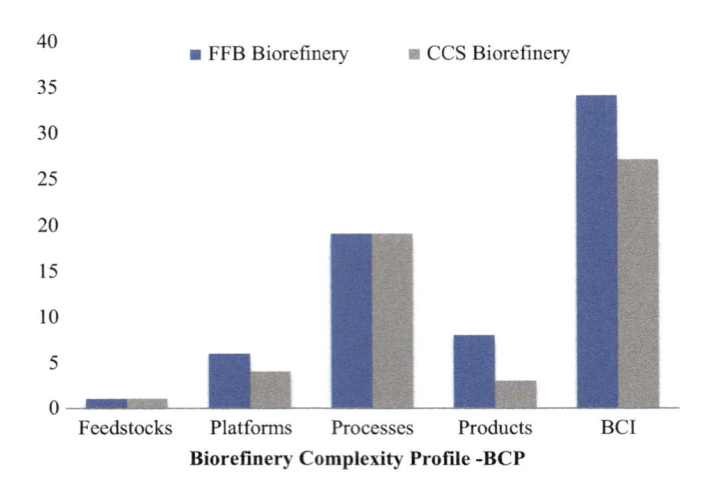

FIGURE 8.9
Biorefinery complexity profiles of FFB and CCS biorefineries.

and BCP of FFB and CCS biorefineries in order to compare the results obtained. The FFB biorefinery that considers the production of biodiesel, hydrogen, butanol, ethanol, acetone, and electricity presents a BCP higher than the biorefinery from CCS that considers the production of furfural, ethanol, and HMF. The BCP of FFB biorefinery is directly influenced by the number of platforms and products. This biorefinery is composed of four platforms such as oil, glycerol, C6 sugars, and lignin, and the biorefinery from CCS considers the platforms of C5 sugars and C6 sugars. The FFB biorefinery takes the residues of biodiesel production (i.e., glycerol) as a substrate to produce ethanol. In addition, it considers the distribution of C6 sugars platform as a substrate for ABE fermentation and anaerobic digestion for obtaining acetone, butanol, ethanol, and hydrogen, respectively. The configuration of the processes explained before gives special characteristics that are reflected in the FCI for each feature of the biorefinery from FFB.

Table 8.4 indicates the mass and energy indices obtained for FFB and CCS biorefineries. In the FFB biorefinery, the main products are biodiesel, hydrogen, butanol, and ethanol; acetone is the coproduct; and electricity is the by-product. The MI_B indicates that 0.25 kg of products can be obtained from 1 kg of raw material. FFB biorefinery includes processes that require high amounts of water for hydrolysis (acid and enzymatic) and fermentations. For this reason, the MI_B^w is considerably low. The flows of required water are higher than the flows of products. The MI_B^r is affected directly because the process of biodiesel needs sodium hydroxide, sulfuric acid, and methanol as the catalyst and reagents, and the process of butanol needs dodecanol to

TABLE 8.4

Mass, Energy, and Economic Indices for FFB and CCS Biorefineries

Mass and Energy Indices					
Biorefinery	MI_B	MI_B^w	MI_B^r	$MI_B^{cop-byp}$	EnI_B	$EnI_{B-equip}{}^+$
FFB	0.25	0.02	0.20	0.01 0.06*	0.641	5.62
CCS (5)	0.31	0.02	0.21	–	0.37	39.70
CCS (25)	0.30	0.02	0.20	–	0.35	41.53
CCS (50)	0.30	0.02	0.20	–	0.36	40.59
CCS (100)	0.31	0.02	0.21	–	0.37	39.67

Economic indices			
Biorefinery	pc (product, coproduct, by-product) (USD/kg)	sp (product, coproduct, by-product) (USD/kg)	EcI_{B1}	EcI_{B2}
FFB	Biodiesel: 0.34 Hydrogen: 9.59 Butanol: 5.96 Ethanol: 6.98 Acetone: 5.69 Electricity: 0.7	Biodiesel: 1.52 Hydrogen: 12.0 Butanol: 3.0 Ethanol: 1.04 Acetone: 1.36 Electricity: 0.14	1.29	1.49
CCS (5)	Furfural: 1,99 HMF: 2.12 Ethanol: 1.07	Furfural: 1.7 HMF: 2.0 Ethanol: 1.04	1.09	–
CCS (25)	Furfural: 2.06 HMF: 2.01 Ethanol: 0.97		1.06	–
CCS (50)	Furfural: 1.34 HMF: 1.91 Ethanol: 0.8		0.85	–
CCS (100)	Furfural: 0.88 HMF: 1.66 Ethanol: 0.72		0.69	–

* *kW per kg of FFB.*
+ *MJ per kg of raw material.*
pc: production cost.
sp: sale price.

carry out the extraction of interesting products after the fermentation. For $MI_B^{cop-byp}$, it is necessary to determine two values. The first one refers to the coproduct (i.e., acetone) from which 0.01 kg from 1 kg of raw material can be obtained. This value is significantly low because the ABE fermentation gives low yields. The second one refers to the by-product (i.e., electricity) from which 0.06 kW/kg of FFB can be obtained. However, electricity is generated from the remaining solids of pretreatment, making it an important value. It covers 100% of the electricity demand and remaining 70% of the produced electricity can be sold.

The EnI_B shows a high value that indicates that 64.1% of stored energy in the raw material is transformed into the products and the coproduct. This high value can be due to the nature of the products (biofuels) and their significant higher heating value. In the energy index associated with equipment, to transform 1 kg of raw material in the biorefinery 5.62 MJ are needed.

In the biorefinery from CCS, the processing scales of raw material do not affect the mass and energy indices. The values present a constant behavior when the scale increases. In general terms, the MI_B shows that 0.3 kg of products can be obtained from 1 kg of raw material. The MI_B^r is mainly affected by the demand of sulfuric acid in pretreatment and solvents such as DMSO and toluene in the purification of HMF and furfural, respectively. The $MI_B^{cop-byp}$ is not calculated because coproducts and by-products are not considered in the biorefinery. The EnI_B indicates that the stored energy of products corresponds to 37% of the energy contained in the raw material. The $EnI_{B-equip}$ is considerably high due to the demand of utilities in the purification stages of main products, with equipment as distillation columns and liquid–liquid extractors.

Additionally, Table 8.4 shows the results of the economic indices for the biorefineries from FFB and CCS. In the case of biorefinery from FFB, when the production costs and sale prices of products, coproducts, and by-products are compared, it is possible to see that only the production costs of biodiesel and hydrogen are lower than the sale prices. The EcI_{B1} corresponds to 1.29, indicating that the total production cost of products is 29% over the total sale price. This means that the biorefinery is not profitable from an economic point of view. The EcI_{B2} is higher than EcI_{B1} because the production cost of acetone is significantly high. The acetone production negatively affects the economic viability of the biorefinery.

The economic indices of the biorefinery that considers CCS as raw material show that as the processing scale increases the EcI_{B1} decreases. These results confirm that the processing scale is a determinant factor in the viability of a biorefinery or biotechnological process. When 5 ton/h of CCS are fed, the total production costs of furfural, HMF, and ethanol are 9% of the total sale prices. However, with the largest processing scale, the biorefinery presents profits of 31%.

8.5 Conclusion

The characterization indices are decision elements in the conceptual design of stand-alone processes and biorefineries, promoting a short work time and simplifies complex calculations to find the best process configuration. In this chapter, it is demonstrated that using different characterization indices applied to these types of processes is an efficient and easy way to determine their viability range. The application of these indices to biorefineries allows us to make key decisions about specific elements such as the selection of feedstock, products, transformation routes, etc., and to find the biorefinery configuration with the best technical and economic characteristics. The characterization indices allow us to perform an elemental analysis of technical and economic feasibility in the generation of a preliminary screen during the conceptual design of biorefineries. The characterization indices are presented as a practical tool that involves easy and understandable calculations.

References

1. D. Johnston, "Complexity index indicates refinery capability, value," *Oil Gas J.*, vol. 94, no. 12, 1996.
2. G. Jungmeier, H. Jørgensen, and E. De Jong, "The biorefinery complexity index," 2014, pp. 1–36.
3. C. A. Cardona, O. J. Sánchez, and L. F. Gutierrez, "Chapter 3: Feedstocks for fuel ethanol production," in *Process Synthesis for Fuel Ethanol Production*, pp. 43–75. Boca Raton, FL: CRC Press, 2010.
4. Q. Jin, H. Zhang, L. Yan, L. Qu, and H. Huang, "Kinetic characterization for hemicellulose hydrolysis of corn stover in a dilute acid cycle spray flow-through reactor at moderate conditions," *Biomass and Bioenergy*, vol. 35, pp. 4158–4164, Oct. 2011.
5. S. I. Mussatto and I. C. Roberto, "Alternatives for detoxification of diluted-acid lignocellulosic hydrolyzates for use in fermentative processes: A review," *Bioresour. Technol.*, vol. 93, pp. 1–10, May 2004.
6. R. Morales-Rodriguez, K. V. Gernaey, A. S. Meyer, and G. Sin, "A mathematical model for simultaneous saccharification and co-fermentation (SSCF) of C6 and C5 sugars," *Chinese J. Chem. Eng.*, vol. 19, pp. 185–191, Apr. 2011.
7. E. C. Rivera, A. C. Costa, D. I. P. Atala, F. Maugeri, M. R. W. Maciel, and R. M. Filho, "Evaluation of optimization techniques for parameter estimation: Application to ethanol fermentation considering the effect of temperature," *Process Biochem.*, vol. 41, pp. 1682–1687, Jul. 2006.
8. W. W. Pitt, G. L. Haag, and D. D. Lee, "Recovery of ethanol from fermentation broths using selective sorption-desorption," *Biotechnol. Bioeng.*, vol. 25, pp. 123–131, Jan. 1983.
9. I. I. Ahmed and A. K. Gupta, "Sugarcane bagasse gasification: Global reaction mechanism of syngas evolution," *Appl. Energy*, vol. 91, pp. 75–81, Mar. 2012.
10. Y. S. H. Najjar, "Gas turbine cogeneration systems: A review of some novel cycles," *Appl. Therm. Eng.*, vol. 20, pp. 179–197, Feb. 2000.
11. W. Tochampa, S. Sirisansaneeyakul, W. Vanichsriratana, P. Srinophakun, H. H. C. Bakker, and Y. Chisti, "A model of xylitol production by the yeast *Candida mogii*," *Bioprocess Biosyst. Eng.*, vol. 28, no. 3, pp. 175–183, 2005.

12. D. De Faveri, M. Lambri, A. Converti, P. Perego, and M. Del Borghi, "Xylitol recovery by crystallization from synthetic solutions and fermented hemicellulose hydrolyzates," *Chem. Eng. J.*, vol. 90, no. 3, pp. 291–298, 2002.
13. S. Shahhosseini, "Simulation and optimisation of PHB production in fed-batch culture of *Ralstonia eutropha*," *Process Biochem.*, vol. 39, no. 8, pp. 963–969, 2004.
14. J. M. Naranjo, "Design and analysis of the polyhydroxybutyrate (PHB) production from agroindustrial wastes in Colombia," Ph Thesis. Universidad Nacional de Colombia. Departamento de Ingeniería Eléctrica, Electrónica y Computación, 2014.
15. N. J. Gil, "Centro de investigacion de la caña de azucar de Colombia—Cenicaña," *Consultorio tecnologico*, 2009. [Online]. Available: www.cenicana.org/banco_preguntas/tema.php?id_grupo=2&id_tema=24&nombre=Preguntas frecuentes Accessed: May 2017.
16. S. R. Marin Pons & Asociados, "Estudio preliminar para producir etanol de la caña de azúcar en la República Dominicana," 2012.
17. M. V. Aristizábal, V.C. A. García, and C. A. Cardona Alzate, "Integrated production of different types of bioenergy from oil palm through biorefinery concept," *Waste Biomass Valori.*, vol. 7, no. 4, pp. 737–745, 2016.
18. V. Aristizábal Marulanda, Y. Chacón Pérez, and C. A. Cardona Alzate, "Chapter 3: The biorefinery concept for the industrial valorization of coffee processing by-products," in *Handbook of Coffee Processing By-Products: Sustainable Applications*, pp. 63–92, Elsevier, 2016.

9

Key Challenges for Future Development of Biorefineries

One of the real goals of biorefineries during the past few years was the proposal of using the biomass as integrally as possible through sustainable technologies and strategies. Biorefineries are a practical way of achieving real developments in the industry for the production of energy and chemicals under an ideal dream of replacing our today's oil with biomass. During the past few years, biofuel facilities have been adding constantly new processing lines without integral design strategies and possibly repeating the past design and implementation errors in oil refineries (resulting in risky projects, pollution, etc.). Like a fashion, these processing lines from biofuel industry have been integrated in a system called "biorefinery," and many sectors have supported this idea through policies to promote the development of the bio-based economies adopting the biorefinery concept. The design of biorefineries is presented as a relevant topic due to the multiple processing paths that could be available to obtain a set of desirable products.

However, after many scientific efforts in design through well-validated methodologies, the already installed biorefineries are currently not working properly. Some big facilities implemented today as biorefineries are closed or working just as stand-alone processes (biofuel plant), but not through a promising multiproduct biorefinery configuration.

While analyzing biorefineries, for example, those implemented after a specific design in the United States and Italy as well as other cases based on existing sugar factories, it is possible to find that there is plenty to do to achieve efficiency. Different strategic approaches are considered in the cited cases: raw materials with inherent logistic restrictions; technical, economic, and environmental assessments together with social considerations; and finally market restrictions. However, a comprehensive design to really consider all these characteristics in the future is still a missing stage to be solved.

The main challenge should be an overall strategy for future design, analysis, and implementation of new biorefineries with a real sustainability and avoiding to copy the same evolution of risky and polemic oil refineries.

Even if the fossil-based economy is represented today by oil refineries working practically at 100% capacity for a very big and growing market, the biorefineries can be a complement or a solution to possible future scarcities in oil (even if today that is not considered as possible after development of fracking technologies). The common oil fuels for transport using combustion engines (e.g., gasoline, jet fuel), high-value chemicals, and petrochemical raw materials continue to be actually a target for biorefineries but with a challenge in sustainable production, even if it is produced from biomass.

The refineries appeared in the 1940s using just a low percentage of the oil and throwing away (polluting) very complex residues that today are fully considered for high value-added products. Actually, in the refineries, about 85% of mass of refined oil corresponds to fuels, whereas the remaining percentage (15%) corresponds to other products. However, oil refining was developed step by step by adding routes and products depending on the market and needs defined by the society through the history. All the not valorized oil fractions were considered residues to dispose in landfills or open ponds. This means that no design strategies for integral valorization of the oil were proposed at the beginning. Today, and after a big pressure from society thinking about the environmental issues, all fractions of oil are practically used to obtain different products. There are two types of refineries in the world: state and private. All refineries use technologies that are highly predictable and optimized, namely, mature technologies [1]. According to the North American Industry Classification System [2,3], petroleum refineries are defined as establishments primarily engaged in refining crude petroleum into refined petroleum. Current biorefineries include the thermodynamic analysis and measures to improve energy consumption in different units such as crude oil distillation (atmospheric and vacuum), fluid catalytic cracking, catalytic hydrotreating, catalytic reforming, alkylation, hydrogen generation, and others related to catalytic and separation processes.

The biorefineries can be considered as an evolution of the biofuel concept in the world, producing a wide range of substances [4–7]. But, finally, this evolution should be based on the sustainability to be demonstrated after a proper design [8].

9.1 Refineries vs. Biorefineries

From Figures 9.1 and 9.2, it is possible to understand the analogy between refineries and biorefineries. However, at the same time, the differences are very clear. In oil refineries, the location, compositions, technologies, and products are practically defined, and the uncertainties are minimal. However, for the biorefineries, even if the purpose is the same, the uncertainties are very high and these aspects are very relevant. Additionally, the location and the scale are the key factors for the sustainable operation of the facility, and the products should be defined after long heuristics and conceptual design analysis of alternatives.

Oil and biomass as raw materials will define exactly the most important differences during the design according to the homogeneity and maturity in

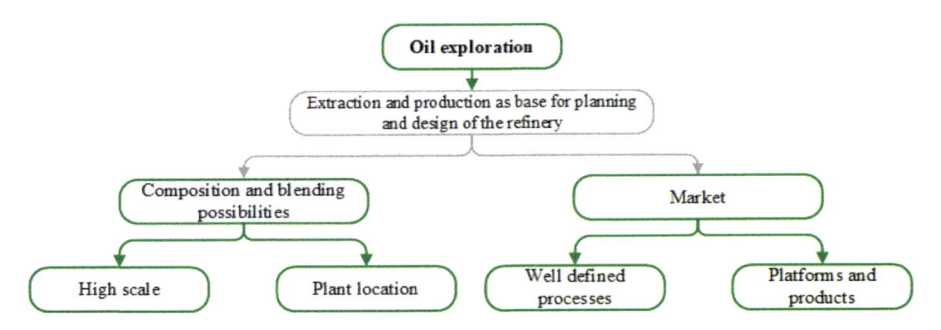

FIGURE 9.1
Refinery design based on very mature technologies, products, and market.

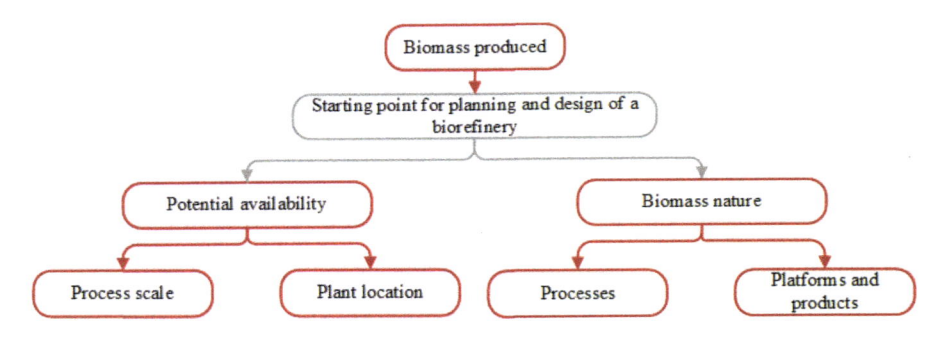

FIGURE 9.2
Biorefinery design based on biomass availability and nature.

technologies for the oil compared with the multicomponent and multiphase biomass. Additionally, the presence of oxygen is very limited in the oil and abundant in biomass. Most of the products obtained at high scale from oil can be resumed in fuels for transport and energy, polyesters and nylon for textiles, polyurethane for adhesives and boards, polymers for bags and packing, glycols for cosmetics, ammonia for fertilizer, synthetic rubbers for tires, and insecticides for human health and agriculture, among hundreds of other products.

The market evolution of petrochemicals defined the sustainability of the oil industry. In the beginning, just some percentage of the oil was used and a huge quantity of residues polluting the environment appeared. However, at that time, no concerns about the environment and the increased demanding for energy (as a result of the progress) made this type of nonsustainable industry possible. As the technologies for fractionation and conversion were developed together with a demand for new products, such as the plastics, the oil as a raw material began to be more and more integrally used, and the residues were reduced to low levels that the industry knows today. However, for biomass as a raw material, this approach cannot be applied. The boom of the biorefineries should be analyzed in a new sustainability context that today is considered as mandatory. Even if biomass is a complex system and the maturity of the technologies are not so high, it is necessary to involve sustainability principles and strategies as discussed in Chapter 5. An overall comparison in terms of sustainability for oil refineries and biorefineries is described in Table 9.1.

TABLE 9.1

Overall Sustainability Principles and Design Strategies to Be Applied in Biorefineries in Comparison with the Oil Refineries

Refinery Design Strategy	Biorefinery Design Strategy	Comments
Refining the same quality of oils and in some cases specific blends. However, in many cases, associated feedstocks such as natural gas were not fully considered.	Integration of feedstocks. As high as integration can be applied, the overall biomass efficiency increases as well as the raw material utilization.	The biorefinery strategy should continue. New refineries now incorporate the use of other feedstocks within the plant as natural and shale gas.
Integrated technologies were the best strategy applied in the past few decades in petrochemical production, e.g., reactive distillation.	Integrated technologies such as fermentation together with saccharification or separation in one unit is the most efficient.	Biorefineries adopted integrated technologies from the petrochemical industry.
The reduction of waste streams was not a key objective in refineries. With time it changed and design strategies such as pinch analysis were involved.	Reduction of waste streams, integrating products with feedstocks in multiprocessing biorefineries.	Biorefinery design included the beginning approaches from oil industry as the pinch analysis.
Refineries design began with stand-alone facilities without a conceptual analysis of the possible and convenient products to be obtained.	Include as many as possible products in the biorefinery. Stand-alone productions could be the worst case. Economic, environmental, and social impacts will always be more positive as the number of products increases [9,10].	Any project based on the biomass should be based on the preliminary or heuristic analysis of the maximal number of products to be obtained after full utilization of biomass.
The refineries for many years did not care about the environment and generally the ecosystems.	Preserve ecosystems and biodiversity. The use of second- and third-generation raw materials reduces the use of other natural sources, which may affect preservation.	Biorefineries should not repeat these errors, and the environmental impacts analysis must have the same importance as the economic assessment, adopting the concept of life cycle analysis.

(Continued)

TABLE 9.1 (*Continued*)

Overall Sustainability Principles and Design Strategies to Be Applied in Biorefineries in Comparison with the Oil Refineries

Refinery Design Strategy	Biorefinery Design Strategy	Comments
Oil industry has the most developed and efficient logistics implementation. However, the real social impact is very low. For example, in terms of distribution, oil is a privilege for some countries.	The supply chain is improved when the biorefinery is working properly in terms of efficiency. This allows increasing the incomes and then redistributing them on the growers or other chain actors.	The logistics and socioeconomic aspects of the biorefineries are more important than those of oil refineries. The supply chain will define the real success of the project considering the huge quantity of biomass that can be distributed along the countries or regions.
In oil refineries, for many years, energy consumption and very complex by-products and residues were not an issue. Energy was considered as very abundant and residual fractions of oil or by-products were just disposed in landfills or open ponds.	Reduction of energy consumption and by-products with low added value should be a challenge during the design.	Energy consumption and by-product utilization or valorization are also an essential part of the design of biorefineries to decide the viability of the project.
Oil industry was the engine for the development of thermodynamics, kinetics, mass–energy transfer, and other topics. From the beginning, this industry designed and developed the most powerful design software.	The use of modern tools and strategies of analysis and evaluation of environmental, technical, and economic impacts is very important. Aspen Plus and other software for specific units and calculations of results in an accurate way are used to predict the behavior of large-scale processes [11,12].	Biorefinery design practically uses all the design software from the oil industry. A specific software is needed to be closer to the real feedstocks and technologies used in biomass conversion.
Even at early stages, oil refineries were very concerned about safety based on the state-of-the-art methodologies for robust calculations.	Design safer processes. The use of chemicals and materials under controllable conditions is a goal of the good simulation to ensure stability of the units.	Biorefinery design uses the same strategies from the oil industry to consider the safety.
Oil industry and petrochemical plants practically never considered past, current, and future scenarios for oil refining (including allocation factors, market products, etc.). However, it	Generally, includes scenarios and sensitivity analysis as a main strategy for their design. Examples demonstrated an increasing understanding of biorefinery possibilities [13].	Scenario and optimization strategies can provide information about the real sustainability of the biorefinery before it is implemented. New oil

(Continued)

TABLE 9.1 (*Continued*)

Overall Sustainability Principles and Design Strategies to Be Applied in Biorefineries in Comparison with the Oil Refineries

Refinery Design Strategy	Biorefinery Design Strategy	Comments
is well explained by the stable market we have today for fuels and polymers, for example		refineries also urgently need this type of analysis based on the real risk that biomass represents for them.
The oil business has established a quota production model to regularize prices. Even at different uncertainties, this model is still working.	The hedging strategies should be considered to avoid unexpected and not desirable changes in prices for raw materials and energy used in the biorefinery. This makes possible to manage financial risks [14].	An organization such as Organization of Countries Exporting Oil (OPEC) could be needed to stabilize the business model in terms of biomass prices.

Source: Based on Moncada et al. (2016) [8].

As a basis for understanding the challenge points of the design and introduction to the biorefinery industry, different multiproduct cases at different levels of maturity or application can be analyzed. Table 9.2 shows just ten examples.

9.2 Coffee Cut-Stem Biorefinery

In coffee production, different residues such as pulp, husk, leaves, and coffee grounds are produced mainly during the coffee harvesting, processing, and final consumption [24]. Other residues that are obtained in high volumes are coffee leaves and coffee cut-stems (CCSs). The latter are obtained when coffee trees are cut to obtain a younger tree with higher coffee productivity. These wastes are seasonally produced and stockpiled for their burning or, in most of the cases, for their auto-degradation.

Currently, the CCSs do not have a well-established use. A techno-economic and environmental analysis for a biorefinery based on CCSs is presented by García et al. (2017) for Colombia as a case study [15]. Different process configurations were calculated and assessed based on Figure 9.3 [15]. In this case, most of the residues were valorized. However, an integral analysis is not applied. The logistics analysis developed at coffee industry level (confidential results and access) demonstrated later that the implementation of this project is practically impossible in the Colombian context based on high inclination fields and the existence of more than 100,000 small farmers, which

TABLE 9.2

Ten Biorefineries Based on Biomass Availability and Nature

Feedstock	Status	Comments	Reference
CCSs to produce ethanol and polyhydroxybutyrate	Proposed, under research	Restricted by logistics. Non-standardized feedstocks	[15]
P. patula bark wood residue to produce ethanol and furfural	Proposed, under research	Low logistics restriction. Standardized feedstock	[13]
Glycerol to acrolein, acetaldehyde, and formaldehyde	Proposed, under research	No logistics restriction. Standardized feedstock	[16]
Blackberry residues for ethanol, xylitol, and antioxidants	Proposed, under research	Logistics restriction. Nonstandardized feedstock. Low-scale possibilities.	[17]
Avocado for producing oil, protein, ethanol, xylitol, phenolic compounds, and energy	Proposed, under research	No logistics restriction. No standardized feedstock. Middle- to high-scale possibilities.	[18]
Palm residues for biodiesel and ethanol	Proposed, under research	No logistics restriction. Standardized feedstock. High-scale possibilities.	[19]
Amazonian fruit for oil, pulp, antioxidants, and flours	Proposed, under research	Logistics restriction. Nonstandardized feedstock. Low-scale possibilities.	[20]
Sugarcane for sugar, ethanol, energy, anthocyanins, and polyhydroxybutyrate	Proposed. Research developed.	No logistics restriction. Standardized feedstock.	[21]
Lignocellulosic residues for ethanol and energy	Crescentino Mature Technology (Proesa). Operational	No logistic restrictions. Standardized feedstocks	[22]
Corn stover, wheat straw for ethanol and energy	Abengoa Hugoton. Mature technology. Operational	No logistic restrictions. Standardized feedstocks	[23]

For detailed description, please refer to the original article in each case.

increase dramatically the transport costs and the raw material price, even if it is a residue. Here, the oil refinery strategies cannot be used to improve these results. However, additional work is needed to demonstrate the lower scale possible to have a feasible project including the logistic restriction. It is important to note that this is mainly a simulation work and no experiments supported the results.

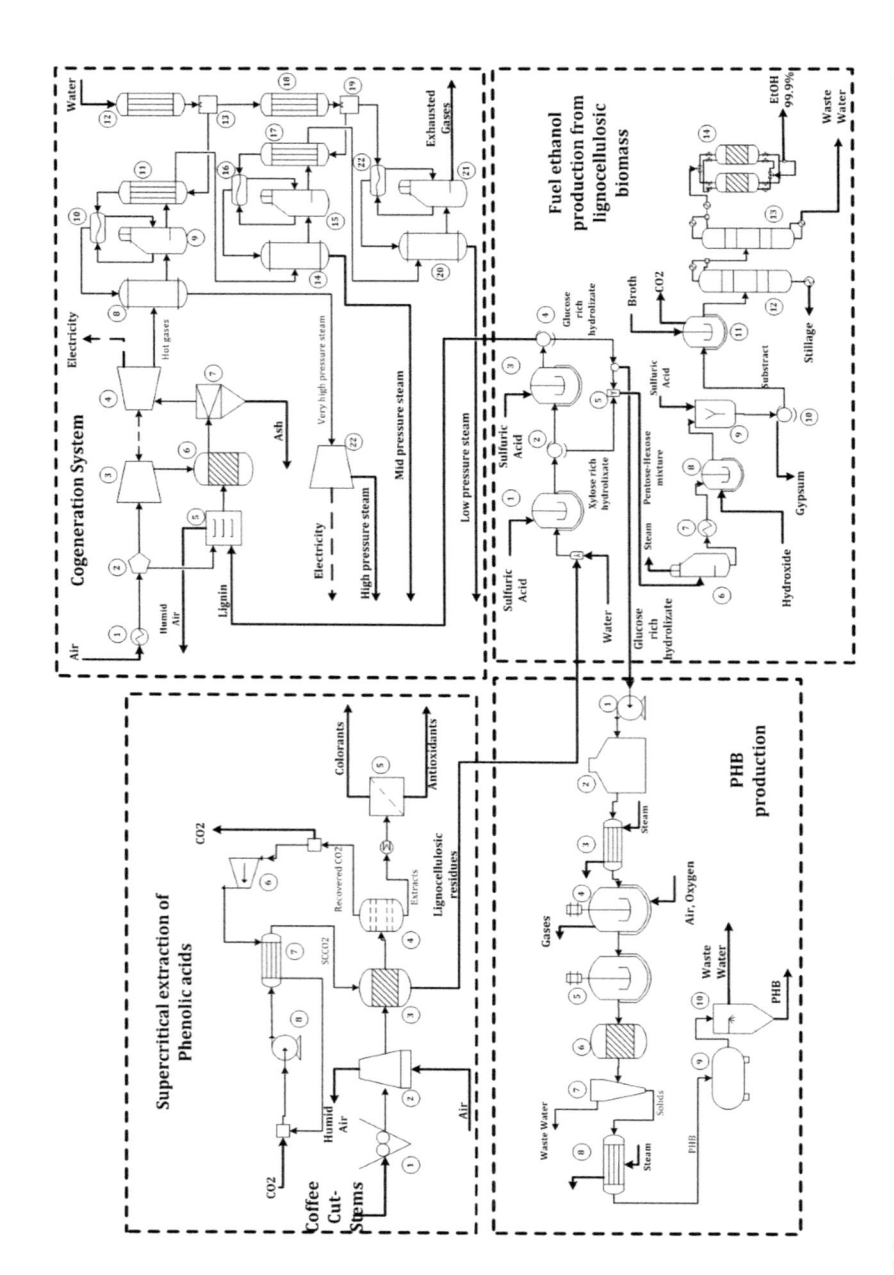

FIGURE 9.3
Biorefinery design based on CCSs.

9.3 *Pinus patula* Biorefinery

Pinus patula is a soft wood widely used in industries producing substantial amounts of residues with just a small portion for specific purposes (e.g., sawdust is used in the cement industry). The idea of a second-generation biorefinery based on lignocellulosic feedstocks (i.e., *P. patula* bark) leads to the formulation and evaluation of biorefining alternatives to obtain different and new products.

Moncada et al. (2016) considered an experimental setup in order to evaluate the technical feasibility of the production of bioethanol and furfural from *P. patula* bark (see Figure 9.4) [13]. To do this, the chemical composition of *P. patula* was determined. In addition, different pretreatment alternatives were tested (i.e., dilute acid, alkali, liquid hot water), including the fermentation of cellulose hydrolysate to produce ethanol (using *Saccharomyces cerevisiae*) and the dehydration of hemicellulose hydrolysate to obtain furfural. Then, the chemical composition of *P. patula* and the yields obtained from the experimental assessment were used to feed a biorefinery simulation (see Figure 9.4) using Aspen Plus to carry out the techno-economic and environmental assessment. Three scenarios were assessed and compared based on different levels of energy integration.

This biorefinery was designed through experiments at laboratory level and simulations with the last tools (Aspen Plus) for process design and analysis. The logistics in this case is not a restriction, but some questions may arise: Is it really the best combination of products? Is there a real and stable market for the so-called fertilizers and animal feed? It is important to note that oil refineries, for example, have a very defined and well-studied market to decide what kind of products will be involved in the plant. It looks today that

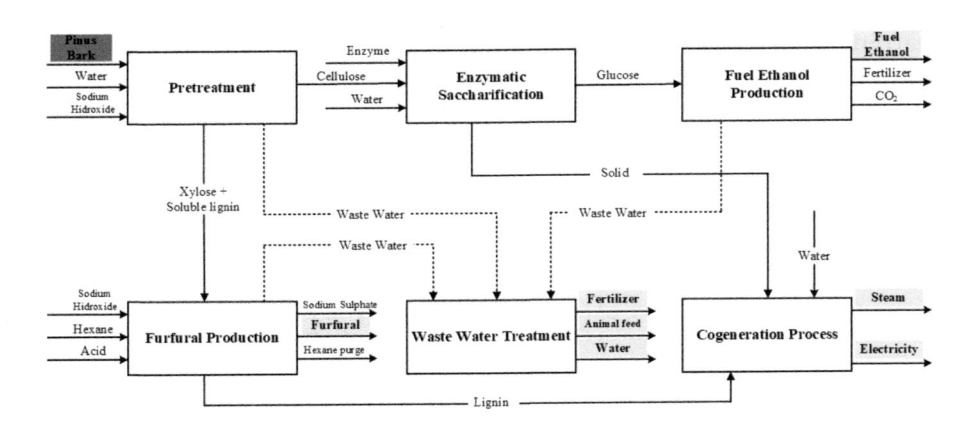

FIGURE 9.4
Biorefinery design based on *P. patula* bark.

most of the biorefineries proposed in the scientific world are very weak in terms of market justification. It could be very dangerous because the errors at this stage can mean a future nonsustainable business.

9.4 Glycerol Biorefinery

Glycerol is a very well-known residue from the biodiesel industry. It can be a very standardized residue without logistics problems. As a raw material for biorefineries, glycerol can be a very good alternative:

- Chemically: versatile molecule, hydroxyl groups (2-prime, 1-s).
- Biochemically: abundant in nature after biodiesel production.
- Both its high functionality and occurrence in nature allow it to be transformed by a chemical route or by fermentation.

Posada et al. (2012) proposed the acrolein, acetaldehyde, and formaldehyde catalytic production as shown in Figure 9.5 [16].

As a result of this very complete work, the total sale price/total production cost ratio for acrolein was about 1.34. This value was higher than the biorefinery case for hydrogen production that reached only 1.09. This case [16] is really interesting for discussion. Even if the biorefinery proposed is very simple, the authors compared conceptually nine chemical and biochemical routes through simulations based on experimental results from literature. One of the most reliable results is to propose acrolein as a better candidate than hydrogen. It is very important to note that hydrogen today is like a tendency in all publications and biorefinery proposals. So, an important learning from this work is related to the not supported choice (i.e., without objective justification) of biorefinery products (very common), which is really

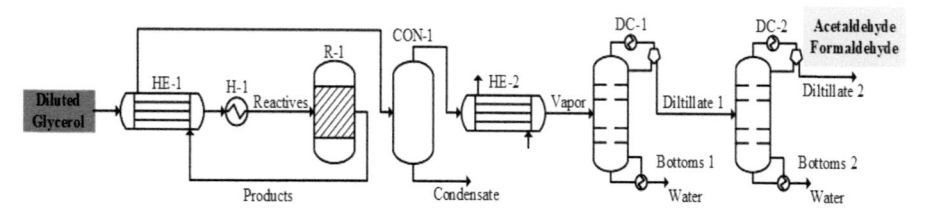

FIGURE 9.5
Biorefinery design based on glycerol. HE-1, heat exchanger 1; H-1, heater; R-1, dehydration reactor; Cond-1, condenser; HE-2, heat exchanger 2; DC-1, distillation column 1; DC-2, distillation column 2.

a risky issue. To compare the different products or a set of products just as a preliminary step at the economic level, it is needed to ensure future sustainable biorefinery projects.

9.5 Blackberry Residues Biorefinery

Andes berry (*Rubus glaucus* Benth.) is a common fruit in many countries, especially in Colombia and other tropical countries. This fruit has an important composition in phenolic compounds, especially anthocyanins that can help reducing carcinogenesis, cancer of epithelial origin, and digestive system cancer among other diseases, using an enhanced fluidity liquid extraction process with CO_2 and ethanol [25]. In 2011, Colombia had a production of 94,000 tons of Andes berry fruits [26], which are used to produce concentrates, jams, and juices, among other products. However, the fruit processing leaves significant amounts of residues, which can be used in the extraction of anthocyanins for chemical, cosmetic, food, and pharmaceutical purposes (Figure 9.6).

Spent blackberry pulp (SBP) is a waste product of blackberry juice extraction with high potential, which is used as a raw material for preparing

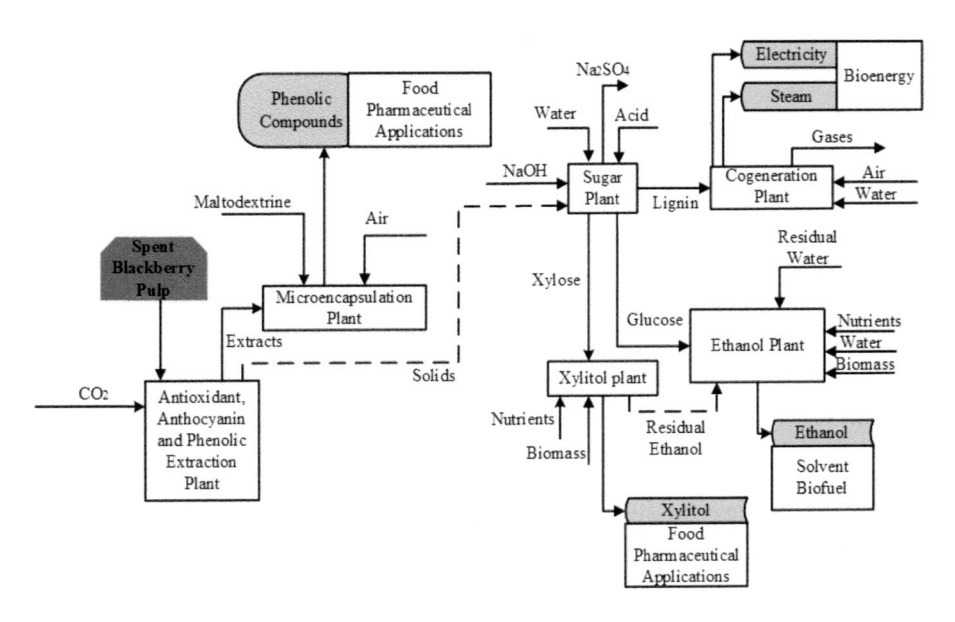

FIGURE 9.6
Biorefinery design based on SBP.

valuable compounds. Dávila et al. (2017) proposed a biorefinery for producing the extract of phenolic compounds, ethanol and xylitol, from SBP [17]. Techno-economic and environmental assessments were developed in order to analyze both total production costs and potential environmental impact of the proposed biorefinery. Four potential scenarios were evaluated: (i) a biorefinery without mass and heat integrations as well as without a cogeneration system, (ii) a biorefinery with heat integration but with neither mass integration nor a cogeneration system, (iii) a biorefinery with mass and heat integrations but without a cogeneration system, and (iv) a biorefinery with mass and energy integrations as well as with a cogeneration system.

The oil refineries in the world are mainly large-scale facilities. Just in the beginning, some small-scale companies existed. The biorefineries can be high scale, but the small-scale cases are not so common given that the energy is usually the main purpose. The only solution to the scale problem is the added value. It is possible to affirm that the products of oil refineries have low added value, and in the case of biorefineries together with biofuels, some high-value-added chemicals can be produced. In the SBP biorefinery analyzed here, the low scale is compensated with very high profits (sale-to-production costs ratio higher than 25). These values can be achieved in practice when molecules such as antioxidants or pharmaceutical products can be obtained. Very high added values in the oil industry are something impossible today, but the market for any value-added product is also very limited.

9.6 Avocado Biorefinery

Avocado (*Persea americana*) is an important commercial tropical fruit. Avocados are rich in unsaturated fatty acids, fiber, vitamins B and E, and other nutrients. The full utilization of this fruit can be totally associated with the commercial pulp extraction industry. Dávila et al. (2017) proposed a model to obtain xylitol, ethanol, oil, antioxidants, protein, and energy [18]. As a result, a feasible biorefinery was proposed as shown in Figure 9.7. This biorefinery can be considered a model, but a number of uncertainties appear: first, the real scale, possible in practice to include bioethanol, is not explained. Second, an integration tendency is not covered in the model, for example, to produce energy for self-supply or to combine the antioxidants with the oil to increase the prices in the market as a functional product. A typical error from the biorefineries is also repeated here; a number of residues are obtained without any proposal or recommendation for their valorization.

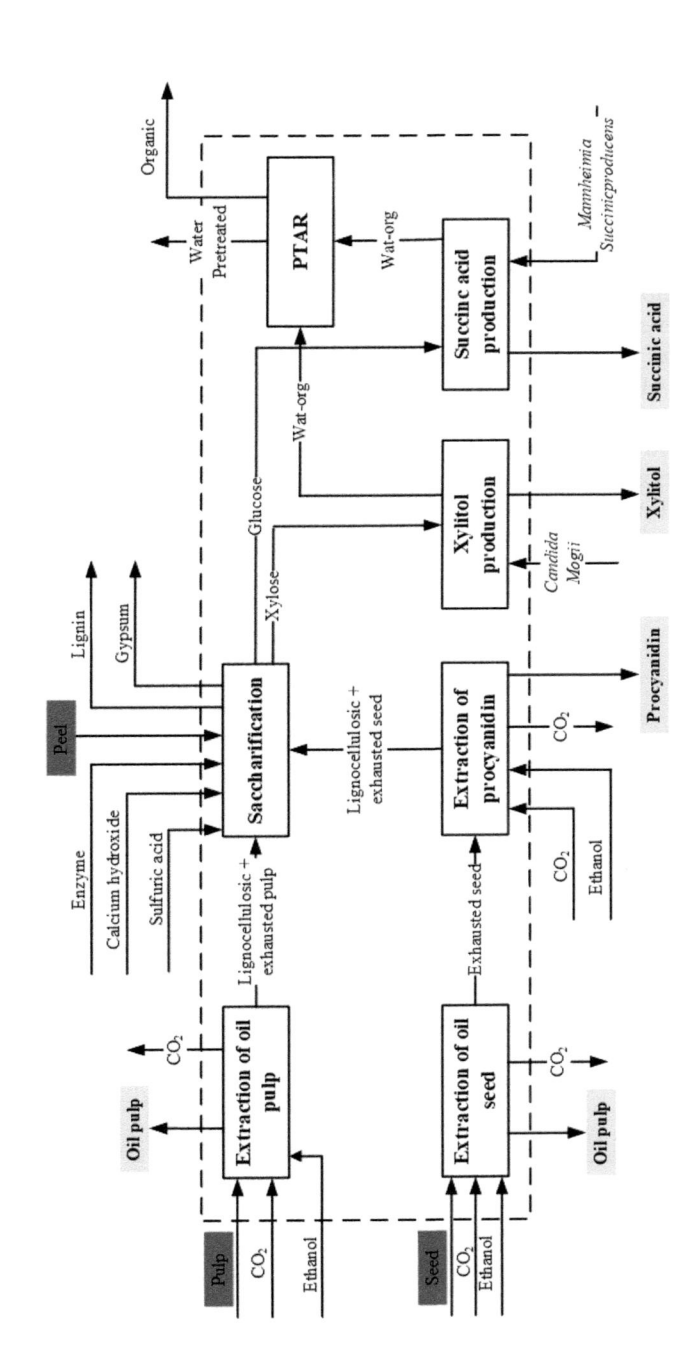

FIGURE 9.7
Avocado biorefinery flow sheet.

9.7 Oil Palm Biorefinery

Rincón et al. (2014) designed two biorefineries based on oil palm, assessed, and compared in order to establish the best route to transform this feedstock into biodiesel, alcohols, and other value-added products [19]. The first integration approach considers the simultaneous production of biodiesel and ethanol from lignocellulosic biomass and glycerol. The second integration approach includes palm oil fractionation, biodiesel production, and biomass-fired cogeneration using a gasification technology. These two integration alternatives were analyzed according to their potential income (total sales/total production cost ratio), and environmental impact (Waste Reduction Algorithm—WAR). The economic and environmental assessment revealed a better global performance for the second integrated approach. In this sense, a biorefinery with a major number of products and low-energy consumption is an important option for the development of oil palm industry. The latter is due to the maximum utilization of this feedstock (Figure 9.8).

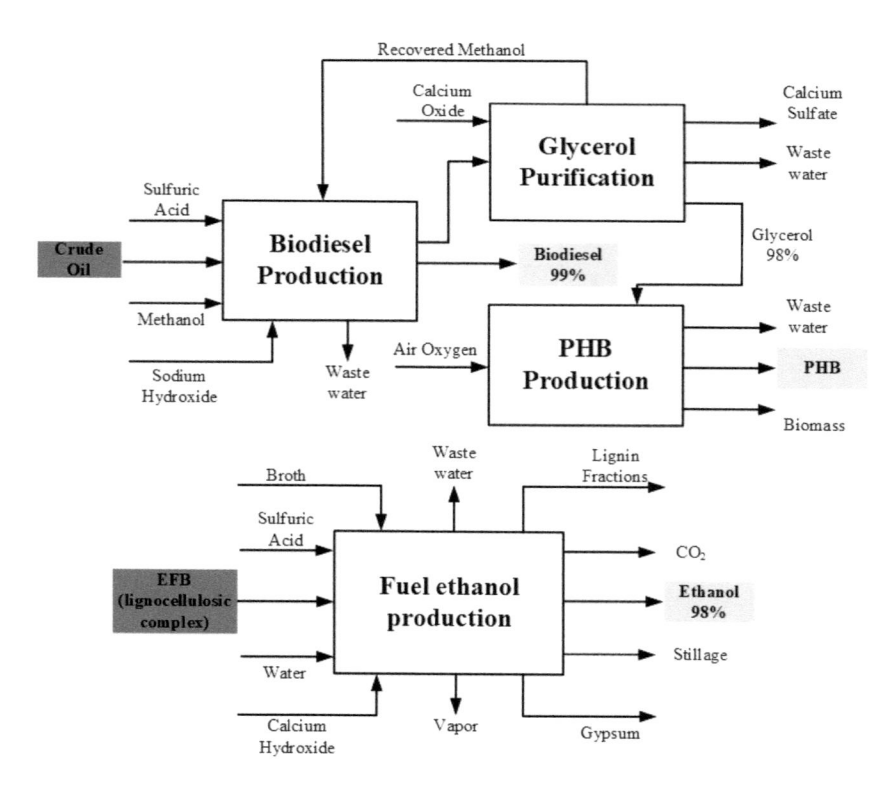

FIGURE 9.8
Oil palm biorefinery.

Palm is an interesting example of total integration of feedstocks in the framework of bioenergy production. The question about the use of methanol (mainly produced in oil refineries) or ethanol (possible to be obtained from the lignocellulosic residues produced in the oil extractions units) demonstrated the competition between oil and biomass industries. Additionally, the logistic problems for energy production in rural areas are also solved through this approach if it is considered that ethanol, biodiesel, and electricity can be produced without dependency on the supply in raw materials from oil refineries.

9.8 Amazonian Fruit Biorefinery

The actual Amazon region is more than any natural region with high biological and ethnic biodiversity. This region is a source of nature riches without exploitation. It is a large watershed that is shared by seven South American countries. However, foremost, it is the scenario of the most serious social, environmental, and territorial conflicts derived to the people movements without organization, exploitation and extraction of resources, and public programs and mistaken regulation or lack of them. The final statements discussed in detail by Correa et al. (1998) involve the justification for new proposals for boosting the economy in this region [27]. González et al. (2016) presented a biorefinery based on an exotic fruit such as Makambo at small scales to produce pasteurized pulp, seed butter, residual cake (it is a paste that should be used as an ingredient in a food industry) as a substitute for cacao, phenolic compounds, biogas, and biofertilizer [20]. This fruit belongs to *Theobroma cacao* (cocoa) and *T. grandiflorum* (copoazú) in the Sterculiaceae family. It grows in different regions of Central and South America. Pulp has been used for direct consumption as food. Seeds are used as a kind of cocoa. Peel is generally disposed as waste. Figure 9.9 shows the biorefinery of the Amazonian fruit.

The analyzed case can be considered as a very small-scale project (2–5 tons a day as maximum). The solutions for small-scale projects should integrate the natural, not very well processed, products (with non-sophisticated technologies) and very high-value-added products such as antioxidants (with very sophisticated technologies). The oil refineries at fractionation steps also use very simple technologies such as distillation with plates only to obtain very high special chemicals in complex catalytic reactors and sophisticated separation units. However, the most important result of this work, where the feasibility of this biorefinery was demonstrated experimentally and theoretically. The possible use of a biorefinery concept at small scale to solve problems in small rural communities was also shown.

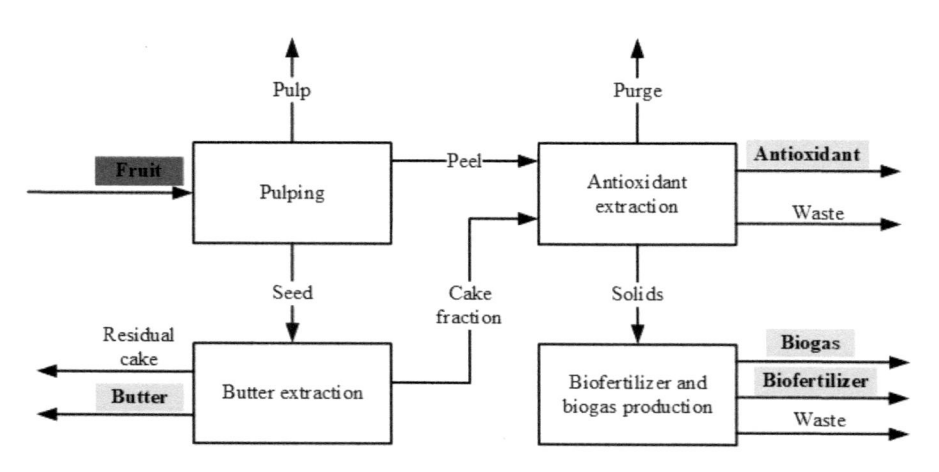

FIGURE 9.9
Amazonian fruit biorefinery: Makambo case.

9.9 Sugarcane Biorefinery

Moncada et al. (2013) presented a techno-economic analysis for a sugarcane biorefinery for the Colombian case [21]. They explained two scenarios for different conversion pathways such as function of feedstock distribution and technologies for sugar, fuel ethanol, polyhydroxybutyric acid (PHB), anthocyanin, and electricity production. These scenarios are compared with the Colombian base case, which simultaneously involves the production of sugar, fuel ethanol, and electricity. They used a simulation procedure in order to evaluate biorefinery schemes for all the scenarios, using Aspen Plus software that included productivity analysis, energy calculations, and economic evaluation for each process configuration. The results showed that the configuration with the best economic, environmental, and social performance is the one that considers fuel ethanol and PHB production from combined cane bagasse and molasses (see Figure 9.10).

One important result to discuss in this work is that a very high profit margin is reached when all the products in an integrated biorefinery are considered. Another important social aspect to consider is the land requirements to ensure the needed feedstock for production. In the case of Colombia, the average crop yield per hectare is 93 tons (but in the main producing regions, it is about 120–140 tons). Therefore, in this project, a high scale is necessary and the sugarcane needed to cover the biorefinery capacity is 1,725,000 tons/year, with a land area requirement of 18,839 hectares. This land requirement represents the 8.6% of the total planted area reported for 2010 [28]. It ensures greater capacity of sugar production to ensure food

Scenario 2

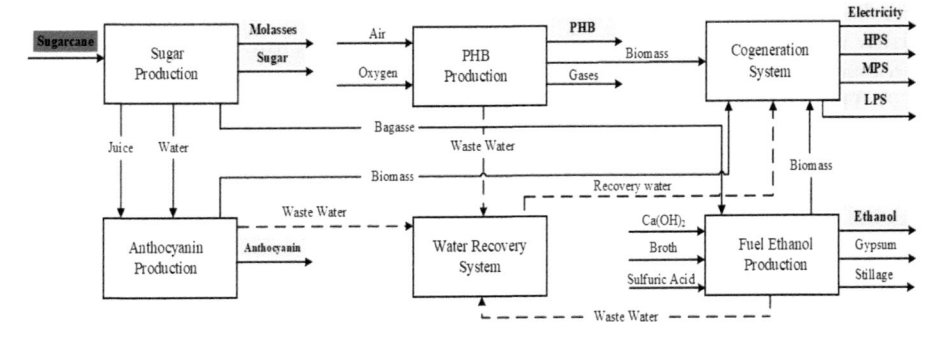

FIGURE 9.10
Scenario 2 (ethanol, PHB, sugar, energy, and anthocyanins as products) for sugarcane biorefinery.

security, larger ethanol production for the oxygenation program, acceptable greenhouse gas emissions, low stillage effluent, and positive social aspects through job generation. PHB production would open new markets to consider second-generation feedstocks, and anthocyanin production would innovate in the national market. The term sequence can be easily used for analyzing other combinations of technologies, feedstock distribution, and inclusion of other products.

One very important topic to discuss in the abovementioned sugarcane biorefinery is the *status quo* that characterizes the oil refineries and the sugarcane sectors. In many refineries, the flexibility for a quick introduction of new process line with new equipments or revamping, is really low. The same existing sugarcane factories are also very conservative, and the addition of new product or conversion lines is practically impossible, even if it is a really new and lucrative business. What was learned after these statements is the importance of flexibility, especially for very big biorefineries, and to be prepared for changes in the market and then in technologies or products needed to ensure a high profitability.

9.10 Beta Renewables Biorefinery

Beta Renewables' commercial-scale production at the Crescentino plant in Italy is one of the most important biomass projects in the world. Owners and designers claim a very low-cost ethanol as a main product. Beta Renewables uses a lignocellulosic biomass technology called Proesa™, which takes non-food-chain crops, or feedstocks, such as giant reed, or agricultural waste material, such as wood residues. The Italy plant was designed to use the region's different agricultural harvest each season, by producing biofuels from wheat straw and corn straw in the summer, rice straw in the winter, and eucalyptus in between the winter and the summer. Proesa combines an enzymatic pretreatment process with fermentation [22]. This process has a shorter residence time than other enzymatic hydrolysis approaches, is acid and alkali free, and has minimal by-products. The parameters are adjustable, providing flexibility in the desired output of C_5 and C_6 sugars and lignin, to be used in the production of chemical intermediates. The technology is protected by 26 patent family applications, 14 of which are public.

The biorefinery was implemented with the promise of producing, biofuels such as ethanol, bio-jet, and butanol, and biochemicals such as fatty alcohols, 1,4-butanediol, farnesene, acrylic acid, and succinic acid, among others. Additionally, lignin derivatives such as phenols, xylene, and terephthalic acid were considered. However, to date, this biorefinery is producing 40,000 ton of ethanol as the main product after 4 years of operation, and actually it is not one of the promised products. This last situation is a very bad message for future projects when considering that a multiproduct biorefinery is still a dream for one of the biggest production projects in the world. Additionally, many logistic problems still appear in one of the best developed projects in infrastructure regions.

9.11 Abengoa's Biorefinery in Hugoton

Abengoa Bioenergy Biomass of Kansas (ABBK) is a company of Abengoa Bioenergy that actually operates a biomass-to-ethanol biorefinery located in Hugoton, Kansas. The construction of this commercial-scale biorefinery facility will allow ABBK to use its proprietary technology to produce 25 million gallons/year of ethanol from nearly 350,000 tons of biomass annually, and the residues are combusted along with 300 tons/day of dry biomass material (feedstock) to produce 18–21 MW of electricity [23]. This power makes the entire facility energy efficient and environmentally friendly as authors claim.

However, this project as one of the biggest in the world was shut down after 1 year of being launched. The so-called biorefinery, built by a Spanish company with financing from the U.S. Department of Energy, never lived up to its billing. Construction of Abengoa's biorefinery finished in mid-August 2014. In late November 2015, Abengoa SA, the parent company of Abengoa Bioenergy, filed for creditor protection in Spain, prompting fears it might default on nearly $10 billion of debt. Then in December, the company shut down the Hugoton plant. Abengoa received a $132.4 million loan guarantee and a $97 million grant through the Department of Energy to support the construction of the Hugoton facility. According to the Department of Energy's website, Abengoa fully paid back its loan guarantee in March.

This is a very good case for further analysis. The financial aspects of biorefineries are still very influenced by the governments and policies. This is a very dangerous support that in this case demonstrates that the sustainability of the projects at every scale should be considered rigorously and the risks accounted before any nice project will be developed.

9.12 Conclusions

An increased growth in biorefinery projects at different scales during the next years is expected. However, it can be concluded that the maturity of biorefineries is still very low, even when very high developed and mature technologies are applied for ethanol or electricity production as the starting point for these biorefinery operations. When oil refineries are compared to biorefineries in terms of evolution and design strategies, the tendency is to repeat most of the errors from the past due to the non-inclusion of the very important aspects: an integral use of raw materials, residues valorization during and not after the design and operation, logistics analysis as well as financial stability. The last just means that the biorefineries from the beginning should be designed under sustainability context.

References

1. IEA Bioenergy Task42, "Biorefineries: Adding value to the sustainable utilisation of biomass," IEA Bioenergy, vol. 1, pp. 1–16, 2009.
2. L. R. Lynd, C. Wyman, M. Laser, D. Johnson, and R. Landucci, National Renewable Energy Laboratory – NREL, "Strategic biorefinery analysis: Analysis of biorefineries," 2005, pp. 1–40.
3. U.S. Energy Information Administration, "Annual energy outlook 2015," *Office of Integrated and International Energy Analysis*, 2015.
4. J. Clark and F. Deswarte, "The biorefinery concept—An integrated approach," in *Introduction to Chemicals from Biomass*, 2nd ed., pp. 1–29. Wiley, 2008.
5. F. Cherubini et al., "Toward a common classification approach for biorefinery systems," *Biofuels, Bioprod. Biorefining*, vol. 3, no. 5, pp. 534–546, 2009.
6. E. de Jong, A. Higson, P. Walsh, and M. Wellisch, "Task 42 biobased chemicals—Value added products from biorefineries," *A Rep. Prep. IEA Bioenergy-Task*, p. 36, 2011.
7. R. van Ree and A. van Zeeland, "IEA bioenergy—Task42 biorefining," 2014, pp. 1–66.

8. J. Moncada, M. V. Aristizábal, and C. A. Cardona, "Design strategies for sustainable biorefineries," *Biochem. Eng. J.*, vol. 116, pp. 122–134, 2016.

9. J. Moncada, J. A. Tamayo, and C. A. Cardona, "Integrating first, second, and third generation biorefineries: Incorporating microalgae into the sugarcane biorefinery," *Chem. Eng. Sci.*, vol. 118, pp. 126–140, 2014.

10. J. Moncada, C. A. Cardona, and L. E. Rincón, "Design and analysis of a second and third generation biorefinery: The case of castorbean and microalgae," *Bioresour. Technol.*, vol. 198, pp. 836–43, Dec. 2015.

11. E. Gnansounou, P. Vaskan, and E. R. Pachón, "Comparative techno-economic assessment and LCA of selected integrated sugarcane-based biorefineries," *Bioresour. Technol.*, vol. 196, pp. 364–375, 2015.

12. H. J. Huang, S. Ramaswamy, W. W. Al-Dajani, and U. Tschirner, "Process modeling and analysis of pulp mill-based integrated biorefinery with hemicellulose pre-extraction for ethanol production: a comparative study," *Bioresour. Technol.*, vol. 101, no. 2, pp. 624–31, Jan. 2010.

13. J. Moncada, C. A. Cardona, J. C. Higuita, J. J. Vélez, and F. E. López-Suarez, "Wood residue (*Pinus patula* bark) as an alternative feedstock for producing ethanol and furfural in Colombia: experimental, techno-economic and environmental assessments," *Chem. Eng. Sci.*, vol. 140, pp. 309–318, Feb. 2016.

14. P. R. Stuart and M. M. El-Halwagi, *Integrated Biorefineries: Design Analysis and Optimization* (Green Chemistry and Chemical Engineering). 2012, Eds. Boca Raton, FL: CRC Press, 2012, pp. 1–873.

15. C. A. García, Á. Peña, R. Betancourt, and C. A. Cardona, "Energetic and environmental assessment of thermochemical and biochemical ways for producing energy from agricultural solid residues: Coffee Cut-Stems case," *J. Environ. Manage.*, vol. 216, pp. 1–9, 2017.

16. J. A. Posada, L. E. Rincón, and C. A. Cardona, "Design and analysis of biorefineries based on raw glycerol: addressing the glycerol problem," *Bioresour. Technol.*, vol. 111, pp. 282–293, May 2012.

17. J. A. Dávila, M. Rosenberg, and C. A. Cardona, "A biorefinery for efficient processing and utilization of spent pulp of Colombian Andes Berry (*Rubus glaucus* Benth.): Experimental, techno-economic and environmental assessment," *Bioresour. Technol.*, vol. 223, pp. 227–236, 2017.

18. J. A. Dávila, M. Rosenberg, E. Castro, and C. A. Cardona, "A model biorefinery for avocado (*Persea americana* mill.) processing," *Bioresour. Technol.*, vol. 243, pp. 17–29, 2017.

19. L. E. Rincón, J. Moncada, and C. A. Cardona, "Analysis of potential technological schemes for the development of oil palm industry in Colombia: A biorefinery point of view," *Ind. Crops Prod.*, vol. 52, pp. 457–465, Jan. 2014.

20. A. A. González, J. Moncada, A. Idarraga, M. Rosenberg, and C. A. Cardona, "Potential of the amazonian exotic fruit for biorefineries: The *Theobroma bicolor* (Makambo) case," *Ind. Crops Prod.*, vol. 86, pp. 58–67, 2016.

21. J. Moncada, M. M. El-Halwagi, and C. A. Cardona, "Techno-economic analysis for a sugarcane biorefinery: Colombian case," *Bioresour. Technol.*, vol. 135, pp. 533–543, 2013.

22. Betarenewables, "PROESA™/The scientific research." Available online: http://www.betarenewables.com/en/proesa/the-scientific-research (Accessed: January 2018).

23. Abengoa Hugoton, "Integrated biorefinery for conversion of biomass to ethanol, synthesis gas, and heat," *DOE Bioenergy Technologies Office(BETO) IBR 2015 Project Peer Review*, 2015.

24. C. R. Soccol, "Aplicações da fermentação no estado sólido na valorização de resíduos agroindustriais," *França-Flash Agric.*, vol. 4, pp. 3–4, 1995.

25. I. X. Cerón, J. C. Higuita, and C. A. Cardona, "Design and analysis of antioxidant compounds from Andes Berry fruits (*Rubus glaucus* Benth) using an enhanced-fluidity liquid extraction process with CO_2 and ethanol," *J. Supercrit. Fluids*, vol. 62, pp. 96–101, 2012.

26. Ministerio de Agricultura y Desarrollo Rural, "Anuario estadístico de frutas y hortalizas 2007–2011 y sus calendarios de siembras y cosechas," 2007, Colombia.

27. A. Correa, J. L. Ramos, B. Hernández, F. Franco, W. Ladino, and A. Prada, *Misión Rural, una Perspectiva Regional: Informe Final*. Tercer Mundo, Ed. pp. 1–217. Santa Fé de Bogota : FINAGRO.

28. Asocaña, "Informe anual 2010–2011. Sector Azucarero Colombiano. Asociación de cultivadores de caña de azúcar de Colombia," 2011.

Index